Telecommunication Systems - Principles and Applications of Wireless-Optical Technologies

Edited by Isiaka A. Alimi, Paulo P. Monteiro and António L. Teixeira

Published in London, United Kingdom

IntechOpen

Supporting open minds since 2005

Telecommunication Systems - Principles and Applications of Wireless-Optical Technologies
http://dx.doi.org/10.5772/intechopen.80140
Edited by Isiaka A. Alimi, Paulo P. Monteiro and António L. Teixeira

Contributors
Tomas Horvath, Petr Munster, Josef Vojtech, Kiran Deep Singh, Ojo O. Adedayo, Oluwafemi B. Ilesanmi, Ogunlade M. Adegoke, Ajibade Adedayo, Ali Hadi Abdulwahid, Juan Manuel Velazquez Arcos, Ricardo Teodoro Paez Hernandez, Jaime Granados Samaniego, Tomas David Navarrete Gonzalez, Isiaka A. Alimi, Paulo P. Monteiro, António L. Teixeira, Ana Tavares, Cátia Pinho, Abdelgader M. Abdalla, Teddy Purnamirza, Toms Salgals, Inna Kurbatska, Sandis Spolitis, Vjaceslavs Bobrovs, Girts Ivanovs, Mário Lima, Amer Al-Canaan, Iman Almomani, Mamdouh Alenezi

First published in London, United Kingdom, 2019 by IntechOpen
IntechOpen is the global imprint of INTECHOPEN LIMITED, registered in England and Wales, registration number: 11086078, 7th floor, 10 Lower Thames Street, London, EC3R 6AF, United Kingdom
Printed in Croatia

British Library Cataloguing-in-Publication Data
A catalogue record for this book is available from the British Library

Additional hard and PDF copies can be obtained from orders@intechopen.com

Telecommunication Systems - Principles and Applications of Wireless-Optical Technologies
Edited by Isiaka A. Alimi, Paulo P. Monteiro and António L. Teixeira
p. cm.
Print ISBN 978-1-78984-293-7
Online ISBN 978-1-78984-294-4
eBook (PDF) ISBN 978-1-83962-597-8

For EU product safety concerns:
IN TECH d.o.o., Prolaz Marije Krucifikse Kozulić 3, 51000 Rijeka, Croatia, info@intechopen.com or visit our website at intechopen.com.

Meet the editors

Isiaka A. Alimi obtained a PhD in Telecommunications Engineering from the University of Aveiro, Portugal, in 2018. He was with Federal Radio Corporation of Nigeria from 2004 to 2012. He joined Federal University of Technology, Akure, Nigeria, in 2013. He is currently a researcher at the Instituto de Telecomunicações, Aveiro, Portugal, where he has been participating in various R&D activities. He has published more than fifteen technical papers and six book chapters. He has served as an academic editor for a book and a guest editor of a special issue. His research interests include network security, optical communication networks, microwave photonics, advanced signal processing and their applications for effective management in access networks. Dr. Alimi is a member of the Institute of Electrical and Electronics Engineers (IEEE).

Paulo P. Monteiro obtained a "Licenciatura" in Electronics and Telecommunications Engineering from the University of Aveiro, Portugal, in 1988; an MSc in Electronic Engineering from the University of Wales, UK, in 1990; and a PhD in Electrical Engineering from the University of Aveiro, in 1999. Presently, he is Associate Professor at the University of Aveiro and Researcher at the Instituto de Telecomunicações. His main research interests include optical communication networks, microwave photonics and electronic subsystems. He successfully tutored and co-tutored more than 14 PhDs, having participated in more than 28 projects. He has authored/co-authored more than 18 patent applications, more than 100 papers in journals and 300 conference contributions. Dr. Monteiro is a senior member of the Institute of Electrical and Electronics Engineers (IEEE).

Dr. António L. Teixeira obtained a PhD in Electrical Engineering from the University of Aveiro, Portugal, in 1999. He is a tenured associate professor at the University of Aveiro and presently the dean of the university's doctoral school. He has successfully tutored more than 70 MSc and 14 PhD students, having participated in more than 35 national and international projects. Previously, he worked at Nokia Siemens Networks and Coriant as a standardization expert. Dr. Teixeira has published more than 400 papers (more than 130 in journals), has co-edited and contributed to several books, and holds 11 patents. In 2014 he co-founded PICadvanced, a startup in the field of telecommunications. He has also served as TPC at major conferences and is a member of several scientific organizations.

Contents

Preface

Telecommunication is the act of transmitting information at a considerable distance with the aid of transmission media. Typical telecommunication equipment comprises a set of well-assembled hardware with embedded software to facilitate reliable and efficient communications. For compatibility purposes, telecommunication equipment performance and design should conform to international standards. Globally, telecommunications have been contributing immensely to economic growth on an unprecedented scale. However, with the ever-increasing deployment of innovative wireless technologies and exponential traffic growth, communication systems have been subjected to significant pressures. To address the challenges, modern telecommunication systems have evolved from being economical and efficient homogeneous systems to converged heterogeneous networks, capable of supporting reliable and quality voice, data, video and other information transmission services. This can be attributed to the advancements in microelectronics and cloud computing-based technologies as well as other fields of telecommunication systems as expatiated in this book.

This book is grouped into four parts: "Telecommunication Protocol, Optimization, and Security Frameworks"; "Next-Generation Optical Access Technologies"; "Convergence of Wireless-Optical Networks"; "Advanced Relay and Antenna Systems for Smart Networks."

Part I presents telecommunication protocol, optimization, and security frameworks. It is noteworthy that there are a number of well-defined protocols in telecommunication systems. These protocols help in achieving efficient communication through collections of distinct rules and regulations. Since telecommunication systems entail several areas, multiple protocols can be employed in different communication fields to accomplish a specified goal. A number of the field-related protocols, such as cellular communication protocols, public switched telephone network communication protocols, data networking protocols, and hybrid communication protocols, are well established. So, well-defined protocols can help in preventing signal loss and in optimizing communication system efficiency. Another important factor in telecommunication systems is system security. All these aspects are comprehensively described in Part I.

The exponential growth in bandwidth demand calls for viable means of meeting system requirements. The advancements in optical communication have been notable solutions to the current and future networks. With optical devices, both short and long transmission can be effectively achieved. Part II provides detailed information on the next-generation optical access technologies. It focuses on different passive optical network variants and advanced optical modulation formats that are vital to telecommunication systems for enhanced spectral efficiency and energy efficiency (EE). Apart from the presented optical fiber-based schemes, free-space optical (FSO) is also considered as a cost-effective, flexible and comparatively easier network deployment. Part II also covers a number of spatial light modulator technologies, which are viable for different applications such as FSO, multicore fiber and spatial division multiplexing.

It is remarkable that wireless and optical network convergence is an attractive solution for leveraging the optical network's inherent bandwidth as well as the mobility benefit of wireless connectivity. Apart from being a good approach for realizing high-network penetration with the anticipated ubiquitous feature, it can facilitate achievement of the 5G network-envisioned EE and capacity. In Part III, enabling technologies for effective convergence of wireless-optical networks are presented. In addition, means of exploiting network convergence in cloud/fog computing are discussed.

Part IV focuses on advanced relay and antenna systems for smart networks. In this part, high-gain and miniaturized antennas capable of operating at microwave and millimeter wave bands are presented. The antennas offer good features such as low-cost, simplified structure, low footprint and ease of installation. Part IV also covers high-speed directional relay based on wireless sensor network for smart grid protection.

In general, this book presents not only the fundamental features, concepts and the associated evolutions of telecommunication, but also explains the underlying technical principles of state-of-the-art systems with good perceptions into the expectation of the next-generation telecommunication systems. The information presented in the book is clear, easy to follow, concise and comprehensible. It covers both theoretical and practical aspects of system implementation in such a way that makes it suitable for students, researchers and professional engineers. It is a good reference for all prospective readers to keep abreast of the current trends in telecommunication systems and to motivate them towards innovative ideas.

Isiaka A. Alimi, Paulo P. Monteiro and António L. Teixeira
Universidade de Aveiro,
Instituto de Telecomunicações,
Aveiro, Portugal

Acknowledgements

The editors would like acknowledge and appreciate all the collaborators that have contributed to the compilation and completion of this book with recent discoveries in different fields of telecommunication systems. Furthermore, we would like to acknowledge the members of the Optical Communications group at the Instituto de Telecomunicações, Aveiro, Portugal, who have vastly contributed with valuable ideas, comments, recommendations and fruitful discussions. In addition, the editors would like to acknowledge support of the European Regional Development Fund (FEDER), through the Regional Operational Programme of Lisbon (POR LISBOA 2020) and the Competitiveness and Internationalization Operational Programme (COMPETE 2020) of the Portugal 2020 (P2020) framework, under Project 5G (POCI-01-0247-FEDER-024539). Also, we acknowledge support of the FEDER through COMPETE 2020 of the P2020, under the Project Virtual Fiber Box with Nr. 033910 (POCI-01-0247-FEDER-033910)]. The projects have contributed immensely both technically and financially towards Chapters 7 and 8. Likewise, we deeply appreciate the technical support being offered under the project DSPMetroNet (POCI-01- 0145-FEDER-029405). Without the project's support, it would have been challenging for this book to come to fruition.

Isiaka A. Alimi, Paulo P. Monteiro and António L. Teixeira
Universidade de Aveiro,
Instituto de Telecomunicações,
Aveiro, Portugal

Section 1

Telecommunication Protocol, Optimization, and Security Frameworks

Chapter 1

Telecommunications Protocols Fundamentals

Amer Al-Canaan

Abstract

The need for communication amongst people and electrical systems motivated the emergence of a large number of telecommunications protocols. The advances in digital networks and the internet have contributed to the evolution of telecommunications worldwide. The purpose of this chapter is to provide students and researchers with a clear presentation of telecommunications core protocols that are utilised in different research domains including telephony, brain-computer interface (BCI) and voice and digital telecommunications. Indeed, BCI involves different electrical signals, communications concepts and telecommunications protocols. This chapter introduces the reader to the core concepts in communications including analogue and digital telecommunications protocols that are utilised generally in communications and in particular in BCI systems. The topics covered in this chapter include telecommunications protocols, communications media, electrical signals, analogue and digital modulation techniques in digital communications, software-defined radio, overview on 10-Mbps Ethernet protocol and Session Initiation Protocol (SIP).

Keywords: BCI, SDR, protocols, Ethernet, analogue modulation, digital modulation

1. Introduction

Telecommunications protocols play an important role in the advanced modern communication systems that convey information, signals and messages over short and long distances. Telecommunications protocols were developed for data (digital) and voice (analogue) messages.

In a typical brain-computer interface (BCI) [1] application, the electroencephalography (EEG) [2] signals are acquired from the brain, encoded and sent over wireless protocols, such as Bluetooth or Wi-Fi data channels, to a control module. However, in a basic BCI system, signals may be sent through wires between signal acquisition and control modules through a certain serial data communications protocol. BCI is one of several vital engineering domains where researchers and students have to understand and deal with telecommunications protocols.

The need for data communications has inspired researchers and led to the emergence of digital communications, integration of Voice over Internet Protocol (VoIP) or IP telephony with multimedia services offered on IP networks over public switched telephone network (PSTN). Modern telecommunications through VoIP software are common on personal computers and portable devices including smart phones and handheld devices. VoIP systems employ packet switching protocols, which have numerous advantages over circuit switching upon which is based on the traditional PSTN.

VoIP applications for local area network (LAN), wide area networks (WAN), wireless local area network (WLAN) and mobile telephone networks offer better availability, scalability, flexibility, minimum hardware and low cost than PSTN. On the other hand, Internet-related problems such as delay and congestion causing jitter and packet loss are inherent in VoIP.

However, circuit switching is compelling in many applications where real-time, low delay and high QoS are desired, where each customer of modern PSTN profits from dedicated analogue or digital circuits. This implies that a communication channel is reserved during a call or a data session. Due to the limited number of circuits and control units in PSTN, only a fraction of customers can perform simultaneous calls within a switch.

One of the main protocols that has been developed for IP telephony is SIP, which is inspired from establishing and ending a call session and for changing parameters of an established session. The simplicity of SIP and the emergence of Java application interfaces for integrated networks (JAIN)-SIP which is a Java-based API for SIP have reinforced the development and implementation of platform-independent IP telephony services.

In this chapter, core concepts in telecommunications protocols, as well as other related topics including communications media, analogue and digital modulation techniques in digital communications, software-defined radio, overview on 10-Mbps Ethernet protocol and SIP protocol, are presented in an easy and simple style with a number of figures to explain the basic principles of telecommunications protocols.

1.1 Telecommunications core concepts

This section introduces the reader to selected core concepts in telecommunications including telecommunications media and digital encoding.

1.1.1 Twisted pair

Twisted pairs are utilised to carry analogue and digital signals. Depending on distance, analogue signals may be limited to 250 kHz, and digital signals are limited to 10 Mbps for distances around 100 m [3]. At the onset of electrical telecommunication systems, copper was the main transmission medium because of its electrical characteristics such as low resistivity to electric current.

1.1.2 Morse code

The Morse code is a variable-length code, where each character is given a series of dots and dashes. Some letters have one dot and others have one dash. The code length varies from 1 to 5, covering 36 symbols. The telegraph signals were carried using copper twisted pairs. Signal wires are twisted in order to cancel out unwanted noise and reduce the effective inductance of the transmission line. At the sending side, a switch is used to open and close the electric circuit in a certain pattern in order to produce Morse code at the receiving side.

1.1.3 Coaxial cable

A coaxial cable consists of a core wire and a cylindrical shield separated by insulation material. It provides better noise rejection and baud rate over longer distances than the twisted pair. Analogue signal frequency can exceed 500 MHz, and baud rate can reach 500 Mbps depending on distance.

1.1.4 Optical fibre

Optical fibre systems consist of a laser diode transmitter and receiver separated by transparent optical fibre. The signals are transmitted as light pulses that propagate inside the optical fibre. The optical fibre has small diameter and consists of three components: the core (pure glass or plastic), the cladding and the protective cover. The cladding material (glass or plastic) is less optically dense, which allows the light to travel easier through the core. The optical fibre can be used on longer distances with attenuation.

1.1.5 Wireless transmission

Radio and TV broadcasting was made possible through various modulation techniques of electrical signals over different carrier frequencies. For example, the short waves (SW) include frequencies from 3 up to 30 MHz, very high frequencies (VHF) range from 30 to 300 MHz and ultra-high frequency (UHF) cover frequency spectra from 0.3 to 3 GHz. Lower frequencies have longer propagation distances, while higher frequencies suffer from reflections and attenuation over long distances. On the other hand, radio frequency (RF) and high-frequency (HF) transmissions require small antennas since their wavelengths are much shorter.

1.1.6 Microwave transmission

With shorter wavelengths in the range 4–6 GHz, microwave signals travel in straight lines and do not penetrate solid objects. They are affected by clouds, rain and obstacles blocking the line of sight between the transmitter and receiver. Usually parabolic antennas are used for large systems. The received signal is focused at the focal point of the parabola.

1.1.7 Very small aperture terminal (VSAT)

In the 1980s, the very small aperture terminal devices made it possible to telecommunicate, utilising small dish dimensions between remote areas by means of highly directional parabolic antennas [4].

1.1.8 Telephone systems

The microphone in a telephone set converts sound into analogue electric signals that are conveyed traditionally through copper wires and reproduced back at the receiver into sound waves through the speaker. The first telephone systems were analogue, while today's telephone systems are completely digital with tone dialling, voice and data services. Telephone networks have profited from advancements in wireless communications by the implementation of the mobile [5, 6] communications. Old telephone networks were designed mainly to convey voice before the emergence of digital data networks and the Internet.

1.1.9 Analogue and digital signals

Digital signals are characterised by two discrete levels, high and low (1 or 0), while analogue signals have continuous forms. Digital and analogue signals are both utilised in modern telecommunications [7] systems and computer networks. Popular digital codes include American Standard Code for Information Interchange (ASCII) and binary-coded decimal (BCD). ASCII is used in basic character symbols for computer systems, while BCD is mainly used for seven-segment displays.

1.1.10 Non-return to zero (NRZ)

Non-return to zero is the simplest digital encoding as shown in **Figure 1**, where a logic one corresponds to a positive high signal level and the logic zero is simply at ground potential or zero voltage. The NRZ encoding is inconvenient for data transmission specially when data contain a long series of zeros or ones.

1.1.11 Return to zero (RZ)

Return to zero is an improved digital encoding over the NRZ encoding, where logic one signals return to zero as shown in **Figure 2**. The RZ encoding is inconvenient for data transmission when data contain a long series of zeros.

1.1.12 Manchester encoding

To assure reliable transmission of digital data (such as Ethernet and IP), the Manchester encoding (refers to **Figures 3** and **4** with clock signal) is convenient to solve the issue of sending a long series of zeros or ones through a data communication line. The Manchester encoding encodes logic 1 as a transition from level high to low signal, while a 0 is a transition from low to high. The needed bandwidth is twice as the original signal, and there is always a change in the middle of each bit.

An improved version of this encoding is called the differential Manchester encoding, where a 0 causes the signal to change at the start of the interval (refer to **Figure 5**). On the other hand, a 1 causes a change at the end of the interval. A 1 keeps the signal level unchanged as in the previous bit and changes to high at the middle. This is advantageous and permits interchanging the wiring of a differential pair without any issue.

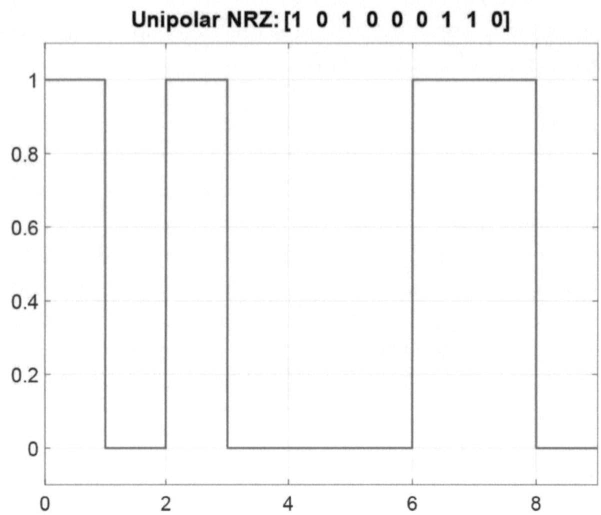

Figure 1.
Unipolar non-return to zero (NRZ).

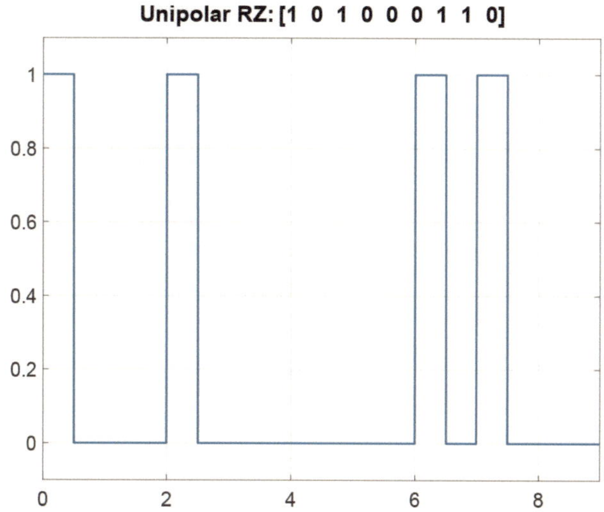

Figure 2.
Unipolar return to zero (RZ).

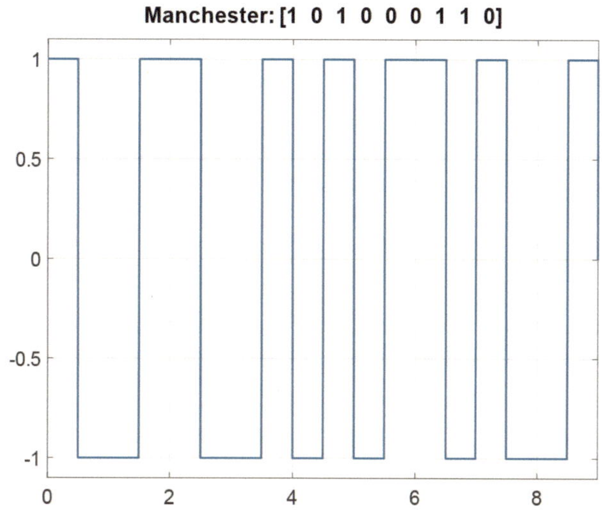

Figure 3.
Manchester encoding.

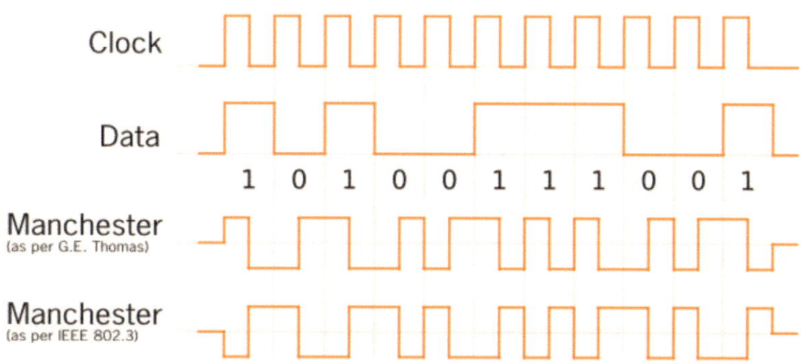

Figure 4.
Manchester encoding example.

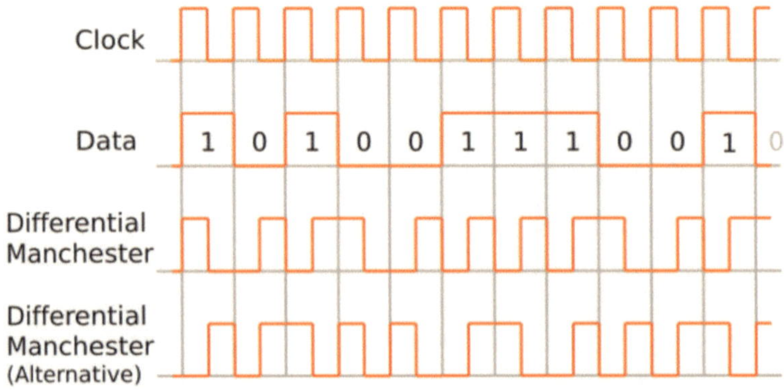

Figure 5.
Manchester differential encoding example.

1.1.13 Shannon's theory

Shannon studied noisy channels, and his theory is based upon the fact that a signal has to have high signal-to-noise (S/N) ratio in order to be successfully distinguished. This influences the maximum bit rate that can be used as follows:

$$Data\ rate\ in\ bps = bandwidth \times log_2(1 + S/N) \tag{1}$$

To increase the data rate, a channel with high S/N should be used. Other means that can increase the bit rate is data compression.

1.1.14 Sampling theory

To convert a continuous signal $x(t)$ into a digital form [8], it is first sampled at equal intervals of time. To be able to reconstruct a sampled signal, $x_\delta(t)$ is defined as

$$x_\delta(t) = \sum_{n=-\infty}^{\infty} x(nT_s)\delta(t - nT_s) \tag{2}$$

The sampling interval T_s is $1/f_s$, where the sampling frequency f_s should be at least twice the highest frequency component f_{max} of the original signal $x(t)$. The frequency $2f_{max}$ is called the Nyquist frequency.

1.1.15 Analogue-to-digital (A/D) conversion

An analogue signal with a given frequency f_1 can be converted into a digital form by sampling it at a constant frequency f_s, where $f_1 < f_s$. A sampled signal has the form of pulses with different amplitudes called pulse amplitude modulation (PAM). The PAM signal is then quantised, and every level is given a binary code number. This process is called pulse-code modulation (PCM). The sampling frequency f_s has to be at least twice as much as the signal frequency being sampled f_1 in order to produce a good approximation of the original signal that can be reproduced and converted back to analogue form. In telephony systems the 8-kHz frequency is used to sample voice that is encoded using 8-bit code. The bit rate in this case is $8000 \times 8 = 64$ kbps. In compact disc (CD) technology, the audio is sampled at 44.1 kHz.

1.1.16 Multiplexing

Multiplexing occurs when data are collected from different sources and are transmitted into one common communication channel. Three types of multiplexing are utilised, namely:

1. Frequency-division multiplexing (FDM). This type of multiplexing employs subcarriers to transmit different message signals.

2. Time-division multiplexing (TDM). This type of multiplexing employs time slots to transmit different message signals.

3. Quadrature multiplexing (QM). This type of multiplexing employs quadrature carriers to transmit different message signals. This type of multiplexing can be distinguished from FDM by the fact that they have overlapped frequency spectra. QM represents double-sideband (DSB) and single-sideband modulations (SSB).

2. Modulation techniques

In the past, digital networks were connected through telephone networks via the modem (modulation/demodulation). Modern telecommunications systems utilise optical fibres that carry many digital channels, which can be translated into voice signals in a telephone by using a codec (coder/decoder). This involves digital-to-analogue (D/A) and analogue-to-digital (A/D) conversions. When a signal $m(t) = A_m \cos\left(2\pi f_m t + \phi_m(t)\right)$ is transmitted, it is normally modulated using a carrier $c(t) = A_c \sin\left(2\pi f_c t + \phi_c(t)\right)$ signal, which can be changed or modulated in amplitude (A_c), phase shift (ϕ_c) or frequency (f_c) [9]. The carrier signal can be generalised as $c(t) = A_c(t)\left[\sin\left(2\pi f_c t + \phi_c(t)\right)\right]$.

2.1 Analogue modulation

To transmit analogue signals over long distances, analogue modulation techniques are used by changing either the amplitude, phase or frequency of analogue signals.

2.1.1 Amplitude modulation (AM)

Amplitude modulation (AM) takes place when $A_c(t)$ is linearly related to the modulating signals (message). In this modulation technique, the carrier frequency is kept constant, and its amplitude is varied according to the amplitude of the transmitted analogue signal as shown in **Figure 6**. An AM signal $y(t)$ is the result of multiplying the message $m(t)$ and carrier $c(t)$ functions. Assuming a sinusoidal carrier signal defined as $c(t) = A_c \sin\left(2\pi f_c t\right)$ is used to modulate the message signal $m(t) = A_m \cos\left(2\pi f_m t + \phi(t)\right)$:

$$\begin{aligned} y(t) &= [1 + m(t)/A_c]c(t) \\ y(t) &= [1 + \mathbf{m} \cos\left(2\pi f_m t + \phi\right)]A_c \sin\left(2\pi f_c t\right) \end{aligned} \tag{3}$$

In the above equation, **m** is the modulation index, which is the ratio of the amplitude of the message signal A_m to the amplitude A_c of the carrier signal.

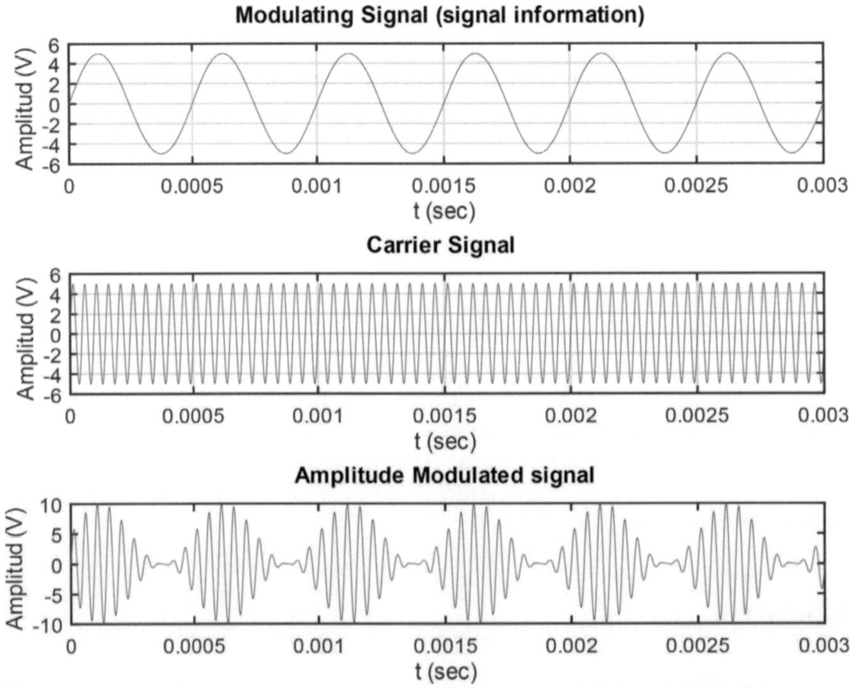

Figure 6.
AM modulation.

To be able to recover the message, **m** should be less than 1, i.e., $1 < \mathbf{m} > 0$. The resulting product function $y(t)$ is composed of three frequencies:

$$y(t) = A_c sin\left(2\pi f_c t\right) + \frac{1}{2}\mathbf{m}A_c\left[sin\left(2\pi\left[f_c + f_m\right]t + \phi\right) + sin\left(2\pi\left[f_c - f_m\right]t - \phi\right)\right] \tag{4}$$

The equation above shows three frequencies:

1. The carrier frequency f_c.

2. The sum of the carrier and modulated frequencies $f_c + f_m + \phi$ with the same phase shift of the message signal.

3. The difference between the carrier and modulated frequencies $f_c - f_m - \phi$ with the negative phase shift of the message signal.

2.1.2 Frequency modulation (FM)

Frequency modulation (FM) takes place when the time derivative of $\phi(t)$ is linearly related to the modulating signal. In this modulation technique, the amplitude of the carrier signal is kept constant, and its frequency is varied according to the amplitude of the transmitted analogue signal as shown in **Figure 7**. Frequency and phase modulations are considered as special cases of angle modulation $s(t) = A_c cos\left(2\pi f_c t + \phi(t)\right)$. The carrier frequency is changed such that the frequency f_c depends on the message signal. Since the frequency is the derivative of the phase, the relation between the input signal and frequency can be written as [9] $\phi'(t) = \mathbf{m_f}m(t)$. The FM signal $y(t)$ can be written as

$$y(t) = A_c\left[\cos\left(2\pi f_c t + \frac{A_m f_\Delta}{f_m}\sin\left(2\pi f_m t\right)\right)\right] \tag{5}$$

In the above equation, A_m is the amplitude of the message signal, f_m is the frequency of the message signal and f_Δ is the maximum frequency that corresponds to the maximum amplitude A_m value. The frequency modulation index $\mathbf{m_f}$ describes the variation in carrier frequency compared [10]:

$$\mathbf{m_f} = \frac{f_\Delta}{f_m} \tag{6}$$

The frequency modulation index can be less than 1 (for narrowband FM) or much greater than 1 (for wideband FM).

2.1.3 Phase modulation (PM)

Phase modulation (PM) takes place when $\phi(t)$ is linearly related to the modulating signal. In this modulation technique, the amplitude of the carrier signal is kept constant, and its phase is varied according to the amplitude of the transmitted analogue signal as shown in **Figure 8**. The phase of the PM signal can be written in terms of the phase modulation index $\mathbf{m_p}$ as $\phi(t) = \mathbf{m_p} m(t)$.

2.2 Digital modulation

Transmission of digital signals involves modulation of amplitude, frequency or phase of carrier signals. The difference between analogue and digital modulation is that in digital modulation, the changes are at discrete intervals. For example, the

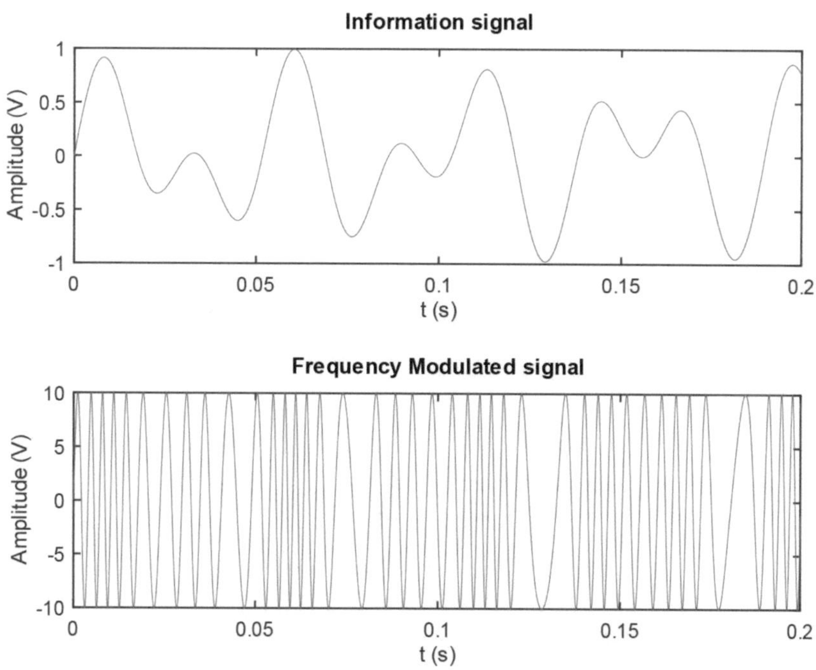

Figure 7.
FM modulation.

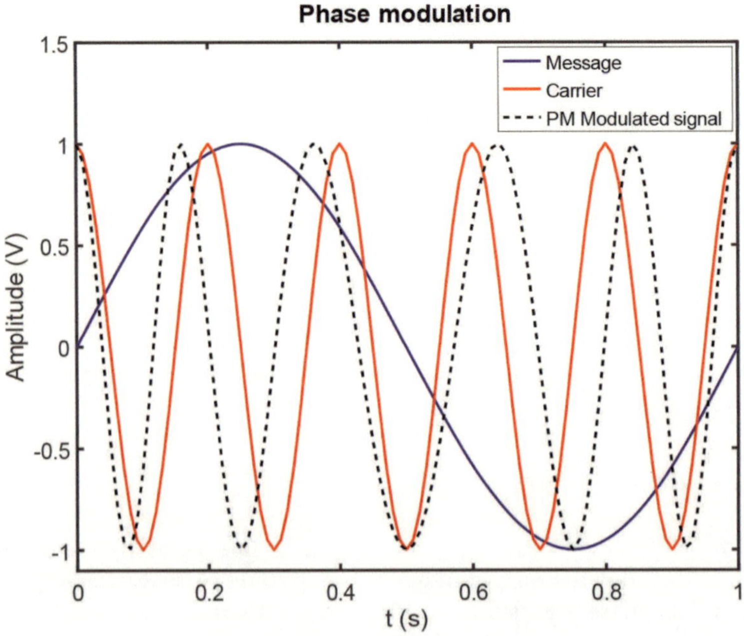

Figure 8.
PM modulation.

amplitude of the carrier signal can be assigned to a maximum value or zero to represent the binary data 1 and 0.

2.2.1 Frequency-shift keying (FSK)

Frequency-shift keying is called also frequency modulation (FM). A bit 0 corresponds to low frequency, and a 1 corresponds to high frequency as shown in **Figure 9**. An FSK signal $s(t)$ can be written as

$$s(t) = \begin{cases} A_c\, cos\left(2\pi\left(f_c + k\right)t\right), & \text{if } bit = 1 \\ A_c\, cos\left(2\pi\left(f_c - k\right)t\right), & \text{if } bit = 0 \end{cases} \quad (7)$$

In the equation above, k is a constant shift in frequency. Obviously, the FSK uses two frequencies ($f_c + k$ and $f_c - k$) for logic 0 and 1, respectively. This type of FSK is called binary FSK (BFSK).

In case k and $3k$ are used to shift the carrier frequency, the resulting FSK signal has four different frequencies and can be utilised to encode the binary codes $00, 01, 10$ *and* 11, as follows:

$$s(t) = \begin{cases} A_c\, cos\left(2\pi\left(f_c + 3k\right)t\right), & \text{if } bits = 00 \\ A_c\, cos\left(2\pi\left(f_c - k\right)t\right), & \text{if } bits = 01 \\ A_c\, cos\left(2\pi\left(f_c + k\right)t\right), & \text{if } bits = 10 \\ A_c\, cos\left(2\pi\left(f_c - 3k\right)t\right), & \text{if } bits = 11 \end{cases} \quad (8)$$

2.2.2 Amplitude-shift keying (ASK)

Amplitude-shift keying is similar to amplitude modulation (AM) as shown in **Figure 10**. Each signal amplitude is assigned to a sequence of bits. If four amplitudes

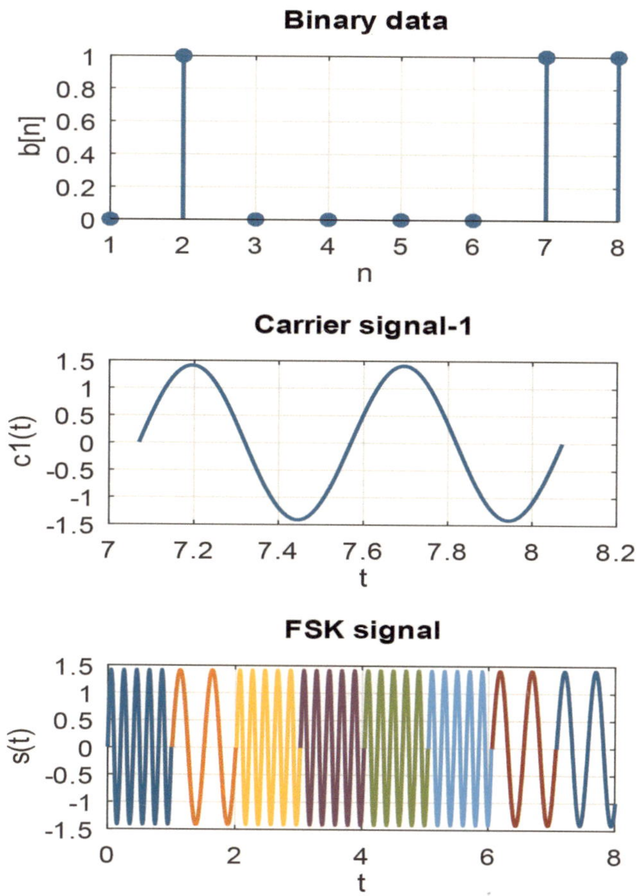

Figure 9.
FSK modulation.

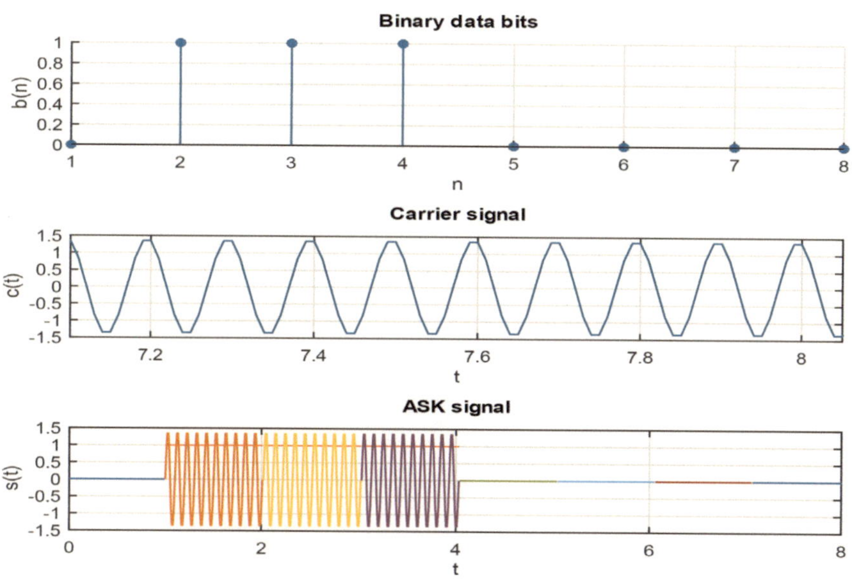

Figure 10.
ASK modulation.

are considered, the following bit code sequences can be defined as 00, 01, 10 and 11. A ASK signal $s(t)$ can be written as

$$s(t) = \begin{cases} A_c \cos\left(2\pi f_c t\right), & \text{if } bit = 1 \\ 0, & \text{if } bit = 0 \end{cases} \quad (9)$$

2.2.3 Phase-shift keying (PSK)

Phase-shift keying (PSK) is also called phase modulation (PM). The signal can have a variable phase as shown in **Figure 11**. If the signal is compared with its predecessor, this technique is called differential phase-shift keying (DPSK). Each phase shift can be assigned to a given binary code [11]. A PSK signal $s(t)$ can be written as

$$s(t) = \begin{cases} A_c \cos\left(2\pi f_c t + \pi\right), & \text{if } bit = 1 \\ A_c \cos\left(2\pi f_c t\right) & \text{if } bit = 0 \end{cases} \quad (10)$$

Since the above equation contains two distinct phases, this type is called binary phase-shift keying (BPSK). If the number of phase variations is increased to 4, the quadrature PSK (QPSK) ca be defined as follows:

$$s(t) = \begin{cases} A_c \cos\left(2\pi f_c t + \frac{\pi}{4}\right), & \text{if } bits = 00 \\ A_c \cos\left(2\pi f_c t + \frac{3\pi}{4}\right), & \text{if } bits = 01 \\ A_c \cos\left(2\pi f_c t + \frac{5\pi}{4}\right), & \text{if } bits = 10 \\ A_c \cos\left(2\pi f_c t + \frac{7\pi}{4}\right), & \text{if } bits = 11 \end{cases} \quad (11)$$

2.2.4 Quadrature amplitude modulation (QAM)

Though the above three approaches can be used with any number of signals, they tend to be difficult to implement due to the fact that special hardware will be needed to distinguish between adjacent amplitudes, phases and frequencies. To overcome this limitation, a combination of bits can be assigned to groups of signals that can be different in amplitude and phase, for example. For example, using signals with two amplitudes and two phase shifts produces four different signals.

2.2.5 Analogue pulse modulation

Pulse modulation can be achieved by modifying either amplitude, width or position of a pulse signal:

- Pulse amplitude modulation (PAM): The PAM signal (as shown in **Figure 12**) is similar to the sampled signal. The pulses in PAM can have a finite width unlike the sampling delta pulses. The PAM-modulated signal $y(t)$ can be written as

$$y(t) = \sum_{k=-\infty}^{+\infty} x(k)\delta(t-k) \quad (12)$$

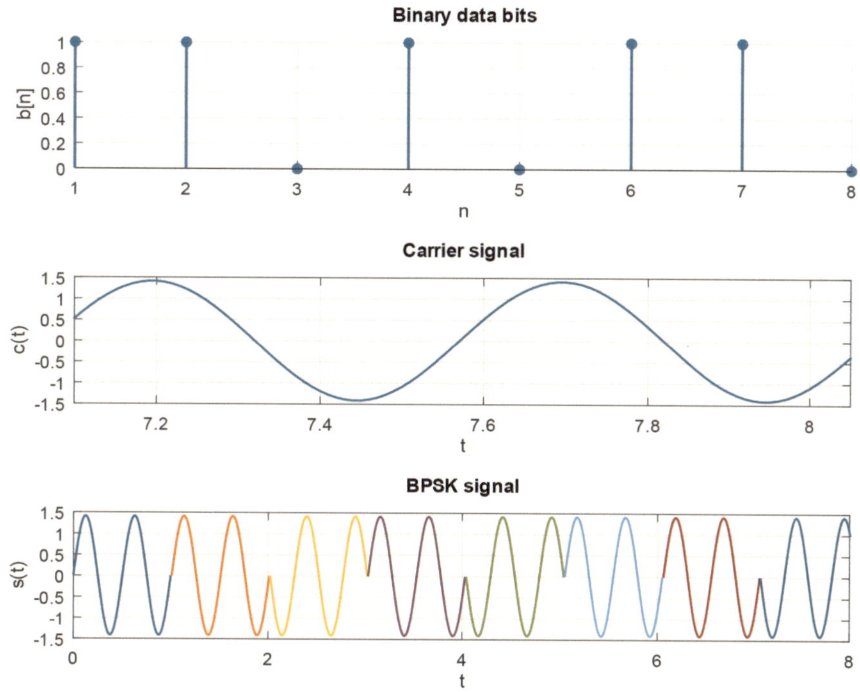

Figure 11.
PSK modulation.

- Pulse width modulation (PWM): In PWM as shown in **Figure 13**, the width of each pulse is related to the modulating signal. This type of modulation is used in DC motor control applications.

- Pulse position modulation (PPM): In PPM as shown in **Figure 14**, the position of each pulse is related to the modulating signal.

2.2.6 Digital pulse modulation

Digital pulse modulation includes two types:

1. PCM: This modulation technique is achieved by sampling the message signal and assigning a digital code (quantisation) to each pulse. The level of the signal is not transmitted; instead the quantised code is assigned according to the available bits for encoding. For example, in 8-bit PCM (with $n = 8$), each level is assigned to a discrete value between 0 and 255. For a signal that has a bandwidth (BW) and a sampling rate of $2BW$, the number of transmitted pulses becomes $2nBW$.

2. Delta modulation: In delta modulation, only the difference between the previous and following codes is sent, as shown in **Figure 15**. For a reference signal $m_s(t)$ and a message signal $m(t)$, the difference $\Delta(t)$ is computed and fed to a pulse generator in order to produce the delta-modulated signal to be transmitted:

$$y(t) = \Delta(t) \sum_{n=-\infty}^{+\infty} \delta(t - nT_s) \quad (13)$$

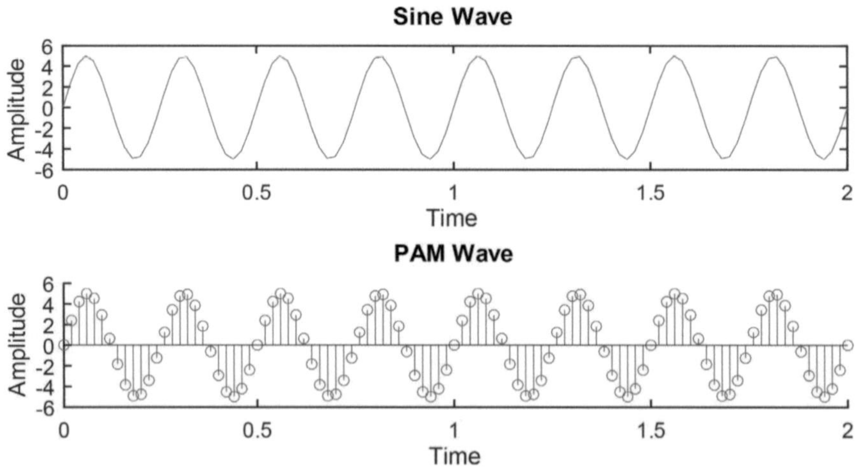

Figure 12.
PAM modulation.

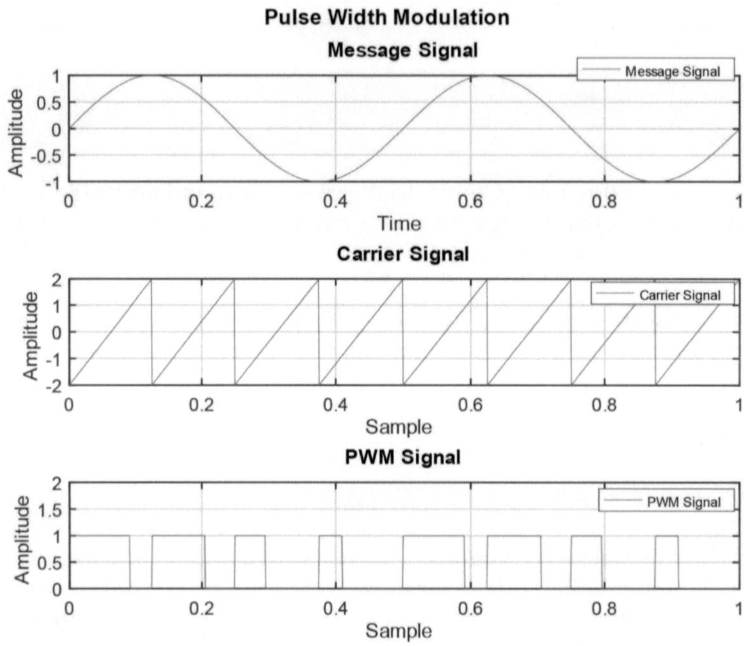

Figure 13.
PWM modulation.

In **Figure 15**, the reference signal $m_s(t)$ is the signal with rectangular edges superimposed on the smooth sine wave message signal. The reference signal is obtained by integrating $y(t)$ as follows [8]:

$$m_s(t) = \sum_{n=-\infty}^{+\infty} \Delta(nT_s) \int^{t} \delta(t - nT_s) d\tau \quad (14)$$

The difference value $\Delta(nT_s)$ is calculated at the nth sampling instant. The reference signal $m_s(t)$ is a stair-step approximation of $m(t)$ as shown in **Figure 15**.

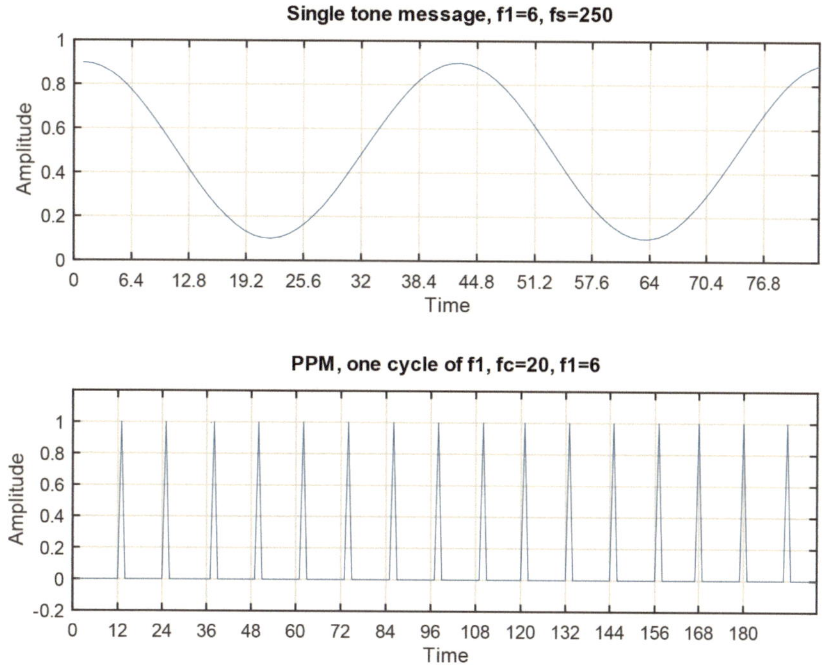

Figure 14.
PPM modulation.

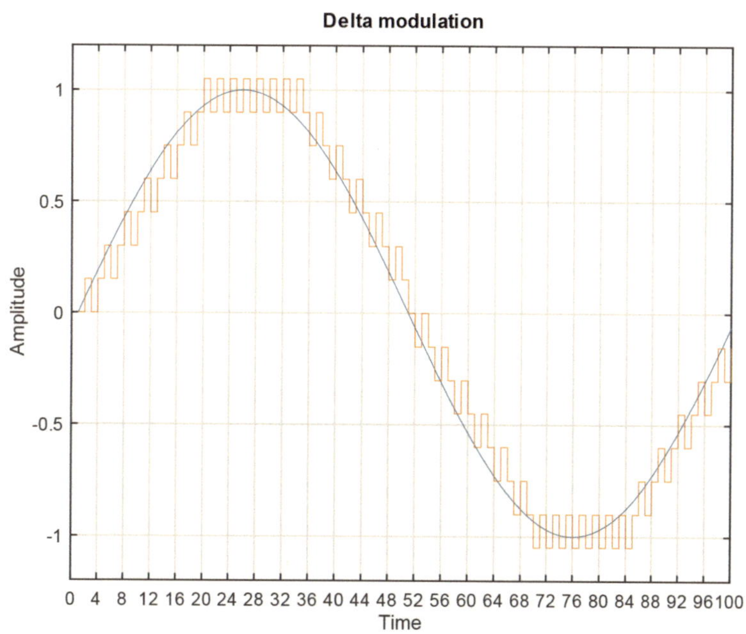

Figure 15.
Delta modulation.

3. Session initiation protocol (SIP)

Modern telephony systems are based upon the Voice over IP (VoIP) protocols, such as SIP, which is a call control and signalling protocol adopted by the 3GPP in order to deliver IP multimedia services [12] to the mobile network [6]. The design of SIP was inspired from HTTP protocol and standardised by the Internet Engineering Task Force (IETF). The purpose of SIP is to enable initiating, terminating interactive call sessions and changing parameters of ongoing sessions. The simplicity of SIP and the emergence of JAIN-SIP [13] have facilitated the development and implementation of platform-independent IP telephony services. Multimedia sessions enable communicating via voice, video and text. SIP messages are either requests or responses and use Session Description Protocol (SDP) in order to determine and negotiate session parameters at either endpoint. SIP supports name mapping and redirection functionalities and, thus, permits user mobility. A typical SIP architecture consists of SIP user agents (UAs) and servers.

4. Software-defined radio (SDR)

Software-defined radio (SDR) [14] is a wireless communication device that employs software to perform most of the operations that are traditionally done by hardware in conventional radio circuits. Similar to the first radio receivers, SDR uses the same hardware for antenna and RF amplifiers. Unlike traditional radios that are based upon hardware to perform modulation and demodulation, software-defined radios are dependent on software to achieve filtering, modulation and demodulation. The IF signal is sampled and converted to digital signal that can be manipulated using software. Common modules between traditional radio and SDR include the antenna and the D/A and A/D converters. Some SDR implementations are freely available using field-programmable gate arrays (FPGA) [15].

5. Overview of 10-Mbps Ethernet

The core protocol of the Internet is the Ethernet protocol, which is based upon serial digital communications. This section provides an overview on the 10-Mbps Ethernet standard. The composition of Ethernet frames (at the MAC sub-layer) and the generation of differential signals at the physical interface (Phy) layer can be implemented on different hardware types as well as FPGA through hardware description language (HDL) code. For 10-Mbps Ethernet, Manchester encoding is utilised, where every bit of information is encoded as a transition from 1 to 0 or from 0 to 1.This is advantageous for the synchronisation between the sender and the receiver and for the recovery of the transmission clock. This encoding method prohibits sending consecutive zeros or ones, which appear as constant DC signal in a conventional RZ encoding. Since every bit of information is composed of two voltage levels, the reference clock is at 20 MHz (double the baud rate).

To identify the beginning of an Ethernet frame, a special pattern of bits is sent, which consists of preamble and a start of frame delimiter (SFD). The preamble and SFD are sent prior to the actual data. The pattern '10' is repeatedly sent, such that a total of 62 bits of 101010 are followed by 11. The last byte (SFD) is 10101011. In hexadecimal, the preamble is 7 bytes of 0x55 followed by a single SFD byte of 0xD5. The first byte that is sent is 0x55, whereas the byte 0x*D*5 is sent last. The leftmost bytes are sent first, of which the rightmost bits (LSB) are sent first. This is why the first byte in the preamble 10101010 is sent from right to left, as 0x55, i.e., the first bit to be transmitted, is 0. Data are usually transferred from an FPGA to the Ethernet port through a physical interface. Taking into consideration the media-independent interface (MII) standard, where the Phy interface communicates nibbles (4 bits) at a time, the SFD 10101011 byte is sent as 0xD and 0x5, since the lower nibble 0xD (in binary, 1011) is sent first starting by 1 (rightmost bit). The reference clocks are 2.5 and 25 MHz for 10-Mbps and 100-Mbps Ethernet, respectively. Reduced MII (RMII) and serial MII (SMII) are two reduced versions of MII, where 2-bit and 1-bit bus widths are used for the Phy, respectively. Compared to the 10-Mbps MII, the gigabit MII (GMII) communicates through 8-bit width bus with a reference clock of 125 MHz. However, the 10-Gbit MII (XGMII) standard deals with 32 bits of data at a time.

Some implementations of Ethernet on FPGA depend upon finite state machines (FSM) programmed in HDL, such as VHDL. Several open-source codes [13] offer Ethernet implementations in VHDL or Verilog.

6. Conclusion

This review chapter contains an overview of telecommunications protocols that are part of modern telecommunications systems. This chapter also provides an overview on analogue and digital signal modulation techniques that are currently used in many research fields including BCI. The researcher in BCI domain as well as the electrical engineering student may find the flow of information smooth and convenient.

The information in this chapter are intended to introduce the reader as well as the researcher in BCI to the core concepts in communications and to analogue and digital telecommunications protocols in an easy-to-follow approach supported with multiple figures and mathematical expression.

The topics covered in this chapter include core concepts in electrical signals, communications, telecommunications protocols as well as other related topics including communications media, analogue and modulation techniques, software-defined radio, 10-Mbps Ethernet protocol and SIP protocol. The topics in this chapter are presented in an easy and simple style with a number of figures to explain the basic principles and fundamentals of telecommunications protocols.

Author details

Amer Al-Canaan
Department of Electrical Engineering, Islamic University of Al-Madinah, Al-Madinah Al-Munnawara, KSA

*Address all correspondence to: amerc@iu.edu.sa

References

[1] Ramadan RA, Vasilakos AV. Brain computer interface: Control signals review. Neurocomputing. 2017;**223**: 26-44

[2] Jiang X, Bian GB, Tian Z. Removal of artifacts from EEG signals: A review. Sensors. 2019;**19**(5)

[3] Shay WA. Understanding Data Communications and Networks. Boston, USA: PWS Publishing Company; 1994

[4] Manohar V, Kovitz JM, Rahmat-Samii Y. Synthesis and analysis of low profile, metal-only stepped parabolic reflector antenna. IEEE Transactions on Antennas and Propagation. June 2018; **66**(6):2788-2798

[5] Sauter M. From GSM to LTE-Advanced Pro and 5G: An Introduction to Mobile Networks and Mobile Broadband. Hoboken, NJ, USA: Wiley; 2017

[6] Lee W. Mobile Cellular Telecommunications: Analog and Digital Systems. Columbus, OH, USA: McGraw Hill Education; 2017

[7] Frenzel LE. Principles of Electronic Communication Systems. 4th ed. Columbus, OH, USA: McGraw-Hill; 2016

[8] Ziemer RE, Tranter WH. Communications Systems, Modulation, and Noise. 6th ed. Hoboken, NJ, USA: John Wiley; 2010

[9] Siva C, Murthy R, Manoj BS. Ad Hoc Wireless Networks. Upper Saddler River, NJ, USA: Printice Hall; 2004

[10] Wikipedia. Frequency Modulation, 2019

[11] Swierczynski P, Fyrbiak M, Koppe P, Paar C. FPGA Trojans through detecting and weakening of cryptographic primitives. IEEE Transactions on IEEE Transactions on Computer-Aided Design of Integrated Circuits and Systems. 2015;**34**(8): 1236-1249

[12] Al-Canaan A, Khoumsi A. Cross-platform approach to advanced IP-telephony services using JAIN-SIP. Journal of Networks. 2010;**5**(7): 8080-8814

[13] Al-Canaan A, Khoumsi A. Advanced IP-Telephony Service Creation using JAIN-SIP API: Crossplatform approach. In: Mosharaka International Conference on Communications, Networking and Information Technology (MIC-CNET 2008); December 2008; Amman, Jordan. pp. 46-51

[14] Machado-Fernandez J. Software defined radio: Basic principles and applications. Revista Facultad de Ingéniera. 2015;**24**(01):79-96

[15] Opencores.org. OpenCores; 2015

Chapter 2

Optimum Efficiency on Broadcasting Communications

Juan Manuel Velazquez Arcos, Ricardo Teodoro Paez Hernandez, Tomas David Navarrete Gonzalez and Jaime Granados Samaniego

Abstract

This chapter is devoted to review a set of new technologies that we have developed and to show how they can improve the process of broadcasting in two principal ways: that is, one of these avoiding the loss of transmission signals due to abrupt changes in sign of the diffraction index and the other, preventing the mutual perturbation between signals generating information leak. In this manner, we propose the join of several of the mentioned technologies to get an optimum efficiency on the process of broadcasting communications showing the theoretical foundations and discussing some experiments that bring us to create the plasma sandwich model and others. Despite our very innovative technology, we underline that a complete recipe must include other currently in use like multiple-input multiple-output (MIMO) simultaneously. We include some mathematical proofs and also give an academic example.

Keywords: wireless communications, optimal broadcasting, information packs, negative refraction index, communication theory, wave propagation through plasma

1. Introduction

Nowadays, one of the most innovative procedures to improve communications is the random scattering of microwave or radio signals that may enhance the amount of information that can be transmitted over a channel. This fact, from a mathematical point of view, is due to the growth of the phase space available for that channel, which provides a more rich mathematical base to define every single signal. In many recent papers, a common subject is the use of a broad range of base functions to span each signal. The hope is that every single collision of the initial signals will be scattered and reaches another phase space region providing additional information, but the increase of phase space involves a more complicated set of describing functions. A multiple scattering of the obstacles enlarges the effective aperture in a time-reversed process for acoustic or electromagnetic signals when they are placed in random manner.

Another current tool is time reversal, or phase conjugation in the frequency domain, where a source at one location transmits sound or electromagnetic waves, which are received at another place, time reversed (or phase conjugated), and retransmitted. The effect is to eliminate noise pollution.

Despite the existence of the mentioned resources and others like multiple-input multiple-output (MIMO), many problems survive, but fortunately, we have proposed some additional ways to improve the broadcasting by diminishing the information loss. Some of our results are based on communication theory and others in the mathematical properties of particular integral equations and their solutions.

Through the present chapter, we introduce for convenience a hypothetical discrete system in order to write finite matrices. But we can certainly extend the validity of our expressions as we will see, even for both discrete and continuum systems provided the involved potentials fulfill very general conditions not discussed in the present work.

In the same manner, because the formalism we have developed for the study of time reversibility refers to acoustic systems, we recall that the scalar wave equation for acoustic signals can be written as:

$$k(\mathbf{r})\frac{\partial^2 f(\mathbf{r},t)}{\partial t^2} = \nabla^2(f(\mathbf{r},t)/\rho(\mathbf{r})) \tag{1}$$

We now describe the quantities appearing in Eq. (1), $\rho(\mathbf{r})$ represents the mass density and $k(\mathbf{r})$ the compressibility of the propagation medium, while $f(\mathbf{r},t)$ is the acoustic signal.

Because the wave equation is of second order in time, we can talk about time reversibility, and then allows solutions, which travel toward the future or the past. An efficient time reversal requires to ensure that the system be ergodic, making possible that the signal may travel both senses in time. To improve focusing, we must describe the signal propagation towards the future or past by means of equations of the same type [18, 22, 27] that is both directions inhomogeneous or both homogeneous. Linearity permits that a signal traveling toward the past can be written with the aid of the integral equation:

$$f(\mathbf{r};T-t) = f^{(\circ)}(\mathbf{r};T-t) + \int_V \int_{-\infty}^{\infty} U^*(\mathbf{r}')G^{(\circ)*}(\mathbf{r}',\mathbf{r};T-t',t)f(\mathbf{r}';T-t')dt'dV' \tag{2}$$

In Eq. (2), $G^{(\circ)*}(\mathbf{r}',\mathbf{r};T-t',t)$ is the free Green function, $U^*(\mathbf{r}')$ depicts the complex dispersion coefficients, and $f(\mathbf{r};T-t)$ is the returning signal that has traveled toward the past. The inhomogeneous term $f^{(\circ)}(\mathbf{r};T-t)$ is known as a sink term and makes both the outgoing and returning equations inhomogeneous integral equations. In Eq. (2), the parameter T represents the time during which the outgoing signal (the one traveling toward the future) is being considered and recording. It is observed experimentally [9] that the time-reversed signal has a definition of a 14th of , the wavelength of the used signal for acoustic signals but this is also true for electromagnetic waves. On several experiments [9, 10], Lerosey, de Rosny, Tourin, and Fink have shown that when such a source term is included, the apparent cross section is increased in two ways: first, the multiple scattering also multiplies the available phase space so when the time is reversed, the information is increased, and second, in the electromagnetic case, the sink term stimulates and triggers the braking of the confinement of the evanescent waves that also raise the information and in consequence the definition to level of about $\lambda/14$. In acoustics, the sink term consists in the operation of the source in reverse order; in the electromagnetic case, the sink term can be implemented with a crest of fine wires around the antennas.

2. Recovering the matrix equations

As we have said above and considering that from a strictly mathematical point of view, both the acoustic and electromagnetic waves achieve the same wave equation type (with a vector version in the electromagnetic case). Then, we can regain, without further ado, the vector matrix formalism [1–7, 11–14] which generalizes the discrete scalar time reversal acoustic model and includes an original model for discrete broadcasting systems that we have called the plasma sandwich model (PSM) [8, 16–18] and we put some associated parameters appeared on it into the named vector matrix formalism (VMF) [8, 20, 24]. But we must underline that is the resonant behavior the one must be considered for increasing efficiency on communications and to achieve extraordinary resolution. To this end, we remember that a three-dimensional version of Eq. (1) can be written as the Fourier transform of an integral generalized homogeneous Fredholm's equation (GHFE) [21–24] for resonances, and does not matter if for acoustic or electromagnetic ones. To analyze the resonant behavior, we must eliminate the inhomogeneous term so we can write the following algebraic equation satisfied by the Fourier transform of the resonant waves:

$$[\mathbf{1}-\eta_R(\omega)\mathbf{K}^{(\circ)}(\omega)]_n^m \mathbf{w}_R^n(\omega)=0 \tag{3}$$

where the kernel $\mathbf{K}^{(\circ)}(\omega)$ is the product of the Fourier transform of the free Green function $\mathbf{G}^{(\circ)}(\omega)$ with the interaction $\boldsymbol{U}$ (without loss of generality we can suppose that $\boldsymbol{U}$ does not depend on ω), so this can be written explicitly as:

$$\left[\mathbf{1}-\eta_R(\omega)\mathbf{G}^{(\circ)}(\omega)\boldsymbol{U}\right]_n^m \mathbf{w}_R^n(\omega)=0 \tag{4}$$

At this point, we must say that we could obtain a transfer matrix description [16–18] instead Eq. (4), but our last equation represents the core of the VMF version. The fact is there are important differences between the two formalisms; for example, VMF makes the time-reversal process easy. Of course, we are moving over a frequency domain and not over a time-dependent one, the former the appropriate in agreement with information theory applications. And certainly, the most important difference is that VMF formalism includes the concept of the resonant solutions.

3. Introducing the PSM parameters

One of the methods we have proposed is based on experiments executed by Xiang-kun Kong, Shao-bin Liu, Hai-feng Zhang, Bo-rui Bian, Hai-ming Li et al. [8] in which they put three layers of plasma joined and alternated with one of them magnetized in the core and the other two unmagnetized in the extremes of the device; when this plasma sandwich is submitted to an external electric potential, it is observed that for a range of values of the external potential, the refraction index is negative [15, 19]. When we analyzed those experiments, we conclude that for this range of the electric potential, the plasma sandwich brakes the confinement of the evanescent waves as occurs in a left-hand material and we proposed a model named the plasma sandwich model for the behavior of the propagation media. Depending on the particular conditions of the propagation media, that is, depending of the values of the plasma sandwich parameters, and for particular conditions of the external electric potential, the propagation media may behave like the plasma sandwich and acquire a negative refraction index. In this section, we introduce the PSM parameters and find the resonant frequencies for a specific problem, underlying that resonant frequencies can be used only to associate an interval of frequencies of a real signal to a device that could be an antenna and not to a single emitted frequency by them; this is because resonant waves are released evanescent waves that vanish in the resource sites and not precisely information carriers. The frequency bands we can build from the resonant frequencies can be considered as convenient highways for the transit of information. Every kernel depends on the response of the media in circumstances that can vary for different time intervals. In this manner, we present an example very easy to work but in which is not relevant the particular behavior of the signal we used to get it. Next, we can find the resonant frequencies for an academic example. First, we choose an appropriate discrete kernel $\mathbf{K}^{(\circ)}(\omega)$, for convenience; in this particular kernel, we do not take into account the three components of the electromagnetic field (usually represented for the indices n and m). However, we propose a system constituted by two emitting antennas. One possible may be written [1, 3–7]:

$$\mathbf{K}^{(\circ)}(\omega)=\begin{pmatrix} \frac{\sin(\omega-\omega_p)\delta}{(\omega-\omega_p)\delta} & -i\frac{\cos(\omega-\omega_p)\delta}{(\omega-\omega_p)\delta} \\ i\frac{\cos(\omega-\omega_p)\delta}{(\omega-\omega_p)\delta} & \frac{\sin(\omega-\omega_p)\delta}{(\omega-\omega_p)\delta} \end{pmatrix} \tag{5}$$

In kernel (5), we have introduced the plasma sandwich model (PSM) parameter δ, which is defined as:

$$\delta \,..\, \kappa \bar{d}_M \tag{6}$$

Definition (6) involves κ with the physical meaning of the wave number of an incident beam that interacts with the magnetic and electric fields in a way that the whole kernel is the expressed in Eq. (5); $\bar{d}_M$ is the average thickness of a plasma-magnetized layer that generates this interaction; parameter ω_p is the average value for the plasma frequency in the magnetized plasma layer which can be written in terms of the local electron concentration in the layer as:

$$\omega_p = \frac{1}{2\pi}\left(\frac{Ne^2}{m\varepsilon_0}\right)^{\frac{1}{2}} \tag{7}$$

In this definition, N is the electron concentration, e is the electronic charge, and ε_0 is the permittivity of vacuum.

It is possible to note that any change in the parameter values gives different broadcasting conditions [5]. PSM suggests that there is not a single stationary set of iterated layers but a bunch of sets evolving in time and in consequence with different effects for each frequency. We must remember that the equation to solve is Eq. (3) where,

$$\mathbf{K}_m^{n(\circ)}\left(\mathbf{r}',\mathbf{r};\omega\right) = \begin{cases} 0 \; if \; \mathbf{r}' = \mathbf{r} \\ U^{nm}\left(\mathbf{r}'\right)\mathbf{G}_{\omega}^{nm(\circ)}\left(\mathbf{r}',\mathbf{r}\right) \; if \; \mathbf{r}' \neq \mathbf{r} \end{cases} \tag{8}$$

The last two ubiquitous conditions to achieve resonance are the vanishing of Fredholm's determinant for Eq. (4), and that Fredholm's eigenvalue λ equals to 1 [6, 11, 22, 23]. The last two conditions give us the expected resonant frequencies for the system constituted by two antennas dependent on the PSM parameters. Now, we must remember that resonances have a special behavior that can be represented by a complex frequency:

$$\omega = \mathrm{K} - i\Lambda \tag{9}$$

The transformation of the evanescent waves for traveling ones is due precisely to the imaginary part Λ. In addition, the relation between ω and the wave number κ is:

$$\kappa = \sqrt{\mu\varepsilon\omega} \tag{10}$$

Substituting expressions (9) and (10) into Eq. (3), we can write the resonance condition as:

$$\Delta\begin{pmatrix} \mathcal{M} & \mathcal{N} \\ \mathcal{N} & -\mathcal{M} \end{pmatrix} = 0 \tag{11}$$

The abbreviated components of the matrix in (11) are explicitly

$$\begin{aligned} \mathcal{M} &= \rho_p[\sin(\rho_p)ch(\gamma_p) - \lambda_p] + \gamma_p sh(\gamma_p)\cos(\rho_p) \\ &+ i[\rho_p sh(\gamma_p)\cos(\rho_p) + \gamma_p\lambda_p] \quad (12) \end{aligned} \tag{12}$$

and

$$\begin{aligned} \mathcal{N} &= \gamma_p\cos(\rho_p)ch(\gamma_p) + \rho_p\sin(\rho_p)sh(\gamma_p) \\ &+ i[\rho_p\cos(\rho_p)ch(\gamma_p) - \gamma_p\sin(\rho_p)sh(\gamma_p)] \end{aligned} \tag{13}$$

In Eqs. (12) and (13), we have used the following definitions:

$$\sigma_M = \bar{d}_M\sqrt{\mu\varepsilon} \tag{14}$$

$$\rho_p = \sigma_M(\mathrm{K}^2 - \Lambda^2 - \omega_p K) \tag{15}$$

$$\gamma_p = \sigma_M\Lambda(\omega_p - 2\mathrm{K}) \tag{16}$$

$$\lambda_p = \lambda(\rho_p^2 + \gamma_p^2) \tag{17}$$

To have an image of the solutions of Eq. (11) (see **Figure 1**), we can make $\mathrm{K} = x$ and $\Lambda = y$ those are the real and imaginary parts of ω, and fix the value for the plasma frequency ω_p so we have the following image:

We obtain for the particular conditions:

$$\mathrm{K} = \Lambda \tag{18}$$

$$\omega_p = 10^6\, Hz \tag{19}$$

The solutions (resonances):

$$x_1 = 5.009 \times 10^5\, Hz \tag{20}$$

$$x_2 = -985.99\, Hz \tag{21}$$

In this case only, x_1 is properly a resonance and x_2 has not physical meaning but maintain their orthogonality properties.

4. Communication theory measurement of information loss

Because we have now a wide vision of the loss of information and we know that this is the reason that the images are not perfect, we can use the results of Shannon, Nyquist, Wiener, Hartley, Hopf [25–29], and other authors that have formulated a measure of the loss of information in communication systems. We support our mathematical results on related works [6, 11, 24, 26, 28], which give us a solid theoretical frame to our present and future papers. Indeed, because the *capacity* of a channel and *entropy* are very close concepts, we can use some of the results we have cited above to answer the problem for TRT and LHM.

Basically, we recall two theorems:

Theorem I.

If the signal and noise are independent and the received signal is the sum of the transmitted signal and the noise, then the rate of transmission is:

$$R = H(y) - H(n) \tag{22}$$

This means that the rate of transmission is the entropy of the received signal less the entropy of the noise. The channel capacity is:

$$C = \underset{P(x)}{Max} H(y) - H(n) \tag{23}$$

Theorem II.

The capacity of a channel of band Θ perturbed by white thermal noise power N when the average transmitter power is limited to P is given by:

$$C = \Theta \log\left(\frac{P+N}{N}\right) \tag{24}$$

In this expression, P is the average power of the transmitted signal and N is the average noise power.

From these two theorems, we make our proposal for a channel where we have lost information in three ways. That is, we have limitations on the maximum frequency Θ (band), the presence of different classes of noise, and on a limited time

T available for a time-reversal process. Then, defining a joint average for the power $Q(n,T)$, the channel capacity is:

$$C_T = \Theta \log \left(\frac{P + Q(n,T)}{Q(n,T)} \right) \tag{25}$$

This remains equal to zero when $P = 0$. The very significant feature of this proposal is the explicit dependence on T, in both the joint average power and the channel capacity, as opposed to the conventional treatment of the signal time duration that is considered as a limit process which tends to infinity. This is a consequence of the explicit form of the Fourier transform of the time-reversed Green function that changes with a factor $e^{i\omega T}$, so even if we are not forced to do so, we can think of it as a parameter that defines the channel. We can think of an arbitrary channel but, when we use it to reverse any signal in time, we follow a different process depending on the time T we decide to fit. Then, we can label the channel with each T as a different one and of course with a different capacity with those corresponding to other values of T. Because of the arguments expressed previously in this work, we can use this measure to the same extent on LHM, ATR, and TRT. For a related discussion of the equivalence of the time-reversal methods and the employment of left-hand materials, we can see ref. [30], and for the use of time reversal on antennas, we can see also ref. [16].

5. An academic example

In order to give an insight into information measurement applied to TR, let us propose that our system behaves like a filter. So, in this particular example, we have no loss if we select $t < T$. We also propose that we have a signal like [12]:

$$\frac{\sin(2\pi\Theta t)}{2\pi\Theta t} \tag{26}$$

And, that we have instead of the incoming signal in Eq. (15) another like [10]

$$\frac{1}{2} \frac{\sin^2(\pi\Theta t)}{(\pi\Theta t)^2} \tag{27}$$

The input function Eq. (26) is a sample of a more general function generated by the sum of a series of shifted functions

$$a \frac{\sin(2\pi\Theta t)}{2\pi\Theta t} \tag{28}$$

where a, the amplitude of the sample is no greater than $\sqrt{S}$ (S is the peak allowed transmitter power).

The channel capacity would be [23] approximately (provided that S/N is small)

$$C_T = \Theta \log \left(\frac{S + Q(n,T)}{Q(n,T)} \right) \tag{29}$$

In the time-reversal process, we have shown that for each Fourier component, we should add a complex exponential factor dependent on T. But we know now that the tool is the same and that only the numerical value of channel capacity C_T

changes. We see how in practice the time-reversal parameter T appears explicitly but also that when we cut the time duration of reversed signal, it is possible to consider them as an additive contribution to $Q(n,T)$. But the form of Eq. (25) suggests a generalized measure of a blend or mix channel capacity when sharing the same band W and differ only by the recording time $T_1, T_2, \cdots, T_n$

$$C_{T_1, T_2, \cdots, T_n} = \Theta \log \left(\frac{S + Q(n, T_1, T_2, \cdots, T_n)}{Q(n, T_1, T_2, \cdots, T_n)} \right) \tag{30}$$

The fact that we are using the same band but different cutting limits $T_1, T_2, \cdots, T_n$ also suggests that we can design an appropriate filter that can distinguish between signals according to the recording time that is we can superpose signals with the same frequency range but with different recording times. In a previous work, we have sketched a filter, but now we give a better-defined device, so we propose (see **Figure 2**) as a hint to get the filter, the following steps for both the transmitter and the receiver:

5.1 Transmitter

First, increase the n frequencies on the unique entrance band $B(\omega_0)$ (that is centered in frequency ω_0) incoming from the inverse of $T_1, T_2, \cdots, T_n$, then the n new top frequencies $\omega_1, \omega_2, \omega_Q$ are used to create n transformed signals with the rule suggested by communication theory and these last signals enter a blender. Then, the mixed signal is taken by a band generator and projected in Q new bands centered at the frequencies $\square_1, \square_2, \square_Q$ (each corresponds to a resonant frequency). Finally, each band enters this signal transmitter.

5.2 Receiver

The Q traveling signals enter the mirror band amplifier, so called because it knows that there are Q resonant frequencies and then can create (or separate the signal in Q resonant bands) Q sub-bands and amplify the signal in each band (at this moment, each band carries a piece of the original n different signals); after this,

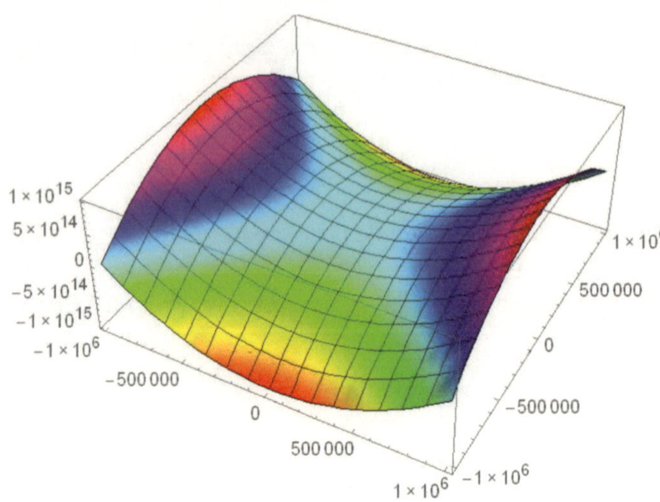

Figure 1.
Image of the solutions of Eq. (11) when the related equation is $987.93(x^2-y^2-10^6) = y(10^6-2x)$.

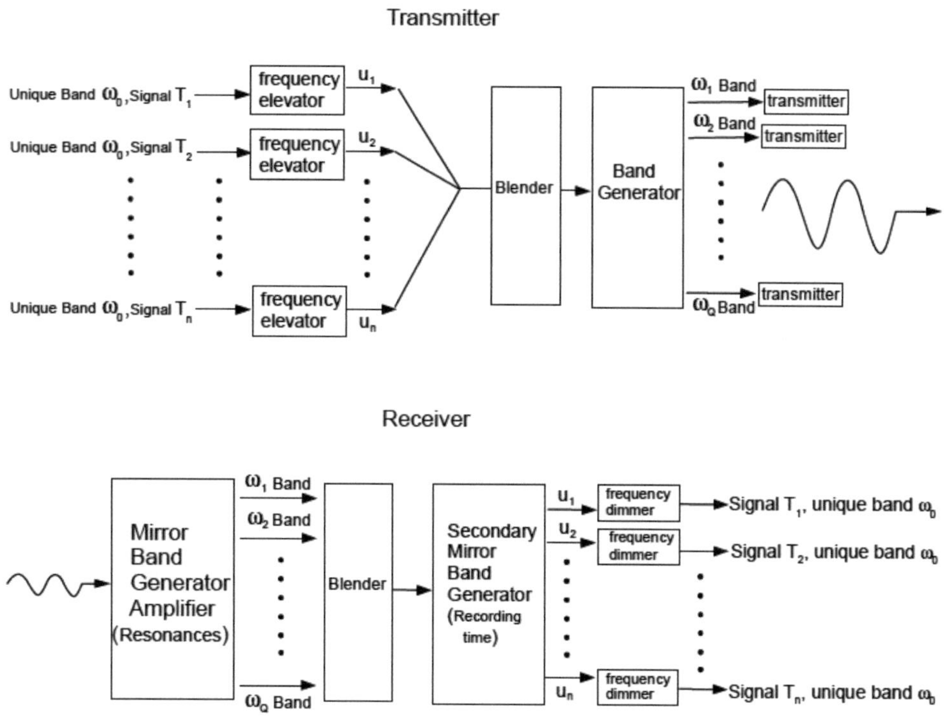

Figure 2.
Flow chart for a proposal device. This can emit and read the blended messages with recording times $T_1, T_2, \cdots, T_n$ *beneath to the unique band* ω_0.

the Q signals are blended and then sending to a secondary mirror band generator which knows that there are n recording times $T_1, T_2, \cdots, T_n$ and because of that it can create n bands with the higher central frequencies $u_1, u_2, \cdots, u_n$ (these last signals could be amplitude-modulated signals) and distribute the blended signal among them. Then, every signal on each band enters a frequency dimmer (the inverse operation performed by the frequency elevators in the transmitter), so we retrieve the n original signals corresponding to the unique band $B(\omega_0)$. For example, in Section 3, we have that the total number of resonances is $Q = 2$, and the two resonant frequencies are $\omega_1 = 5.009 \times 10^5\, Hz$ and $\omega_2 = -985.99\, Hz$.

At this point, it is important to say that the key point on the use of the proposed device is the build of information packs described in another place in order to diminish mutual interference between different signals.

6. Error in time reversing and a related theorem

Based on the equivalence of the TRT and the properties of the Green function, we can trust that any discussion about the interaction of metamaterials with electromagnetic field can be done through this function and simultaneously observe the effect of a time reversal. For this reason, we can now describe the error in terms of the Green function by the hypothesis that LHM can be put to test by forward and backward in time signals and read the results with two points of view: first, the direct effect of the loss of information because of the limited record time T or second, how the negative refraction index helps to preserve information. Now, we can review our previous results and generalize using the kernels, so we can characterize the capacity of a channel in many different circumstances. So, we have made use of the analogies [30] between the TRT and the employment of LHM to propose that we can express the capacity of any of these negative refraction index materials in the same terms or procedures as those of TRT. Also, we can propose an identical description for the channel capacity that is Eq. (24) and its generalization Eqs. (25) and (30). Then, the matrix formalism for discrete systems can be used to characterize the channel capacity of transmission of information in a process of time reversibility using the Fourier transforms of the Green functions (properly we use the kernels with the interaction matrix $\mathbf{V}=\mathbf{1}$) forward and backward. That is, by the first step, the signal transforms like (in the following equations I and F stand for initial and final places):

$$\mathbf{Y}_F=[\mathbf{1}+\mathbf{R}(\omega)]\mathbf{X}_I \tag{31}$$

then in the second step, it returns to the initial place by means of the operation.

$$\mathbf{Z}_I(\omega)=[\mathbf{1}-\mathbf{K}^{(\circ)}(\omega)]\mathbf{Y}_F(\omega) \tag{32}$$

Then, the complete signal trip would be:

$$\mathbf{Z}_I(\omega)=[\mathbf{1}-\mathbf{K}^{(\circ)}(\omega)][\mathbf{1}+\mathbf{R}(\omega)]\mathbf{X}_I(\omega) \tag{33}$$

So that by defining the error in the time-reversing process by:

$$\delta\mathbf{X}_I=\mathbf{X}_I-\mathbf{Z}_I \tag{34}$$

We can write this like:

$$\delta\mathbf{X}_I(\omega)=\mathbf{X}_I(\omega)-[\mathbf{1}-\mathbf{K}^{(\circ)}(\omega)][\mathbf{1}+\mathbf{R}(\omega)]\mathbf{X}_I \tag{35}$$

or

$$\delta\mathbf{X}_I(\omega)=-[\mathbf{R}(\omega)-\mathbf{K}^{(\circ)}(\omega)-\mathbf{K}^{(\circ)}(\omega)\mathbf{R}(\omega)]\mathbf{X}_I(\omega) \tag{36}$$

Eq. (36) is a corollary that shows explicitly the role of both the forward and backward Fourier transforms of the Green function (we have done $\mathbf{V}=\mathbf{1}$ on Eq. (8) for convenience and also for the complete kernels $\mathbf{K}(\omega)$ and $\mathbf{R}(\omega)$). Eq. (36) is very clear about the origin of the errors because we can see, for example, that in the case that the forward and backward Green functions are mathematically one the transpose conjugated of the other for a perfect time reversal (when acting the first on a column vector and on a row vector the other), we get that the error is zero and that the error increases as the differences of both functions also increase. In a very special case, we can then propose that $\mathbf{K}(\omega)$ and $\mathbf{R}(\omega)$ only differ by the factor $e^{i\omega T}$ or $e^{2\pi i\frac{\omega}{\omega_T}}$ when the only source of error is the recording time T, so that we obtain from Eq. (36) that:

$$\delta\mathbf{X}_I(\omega)=-\left[e^{-2\pi i\frac{\omega}{\omega_T}}\mathbf{K}(\omega)-\mathbf{K}^{(\circ)}(\omega)-\mathbf{K}^{(\circ)}(\omega)e^{-2\pi i\frac{\omega}{\omega_T}}\mathbf{K}(\omega)\right]\mathbf{X}_I(\omega) \tag{37}$$

In Eq. (37), the function $e^{-2\pi i \frac{\omega}{\omega_T}}\mathbf{K}(\omega)$ has the form of the Fourier transform of the Green function but with the argument translated by an amount equal to the recording time T that appears explicitly in Eq. (19) that is the Fourier transform of:

$$\mathbf{K}(t-T) \tag{38}$$

But with the time running backward, so, as we will show in a moment, if T is very short, the error will be very huge. On the contrary, if the time goes to infinity, the error will go to zero. Resuming, the new Eqs. (33)–(38), make possible a characterization of the lost information in left-hand materials not only for microwave range, but also for visible frequencies because we have extended recently the time-reversal techniques (see ref. [3, 12]).

Now, we can define:

$$\mathcal{K} = e^{-2\pi i \frac{\omega}{\omega_T}}\mathbf{K} \tag{39}$$

So, we can write Eq. (37) like:

$$\delta\mathbf{X}_I(\omega) = -[\mathcal{K}(\omega) - \mathbf{K}^{(\circ)}(\omega) - \mathbf{K}^{(\circ)}(\omega)\mathcal{K}(\omega)]\mathbf{X}_I(\omega) \tag{40}$$

and because the kernel of the Fourier transform of the generalized inhomogeneous Fredholm's equation (GIFE) satisfies the following integral equations:

$$\mathcal{K} = \mathbf{K}^{(\circ)} + \mathbf{K}^{(\circ)}\mathcal{K} \tag{41}$$

$$\mathcal{K} = \mathcal{K}^{(\circ)} + \mathcal{K}^{(\circ)}\mathcal{K} \tag{42}$$

While Eq. (41) exactly represents the problem with a finite recording time T, Eq. (42) represents a hypothetical problem in which the recording time is infinite.

Substituting Eq. (41) into Eq. (40), we have:

$$\delta\mathbf{X}_I(\omega) = -[\mathcal{K}(\omega) - \mathcal{K}(\omega)]\mathbf{X}_I(\omega) \tag{43}$$

Then, we can suppose that the two kernels in Eq. (40) represent the real and the hypothetical problem described above. Of course, we see that if real conditions approximate the ideal ones, the error is clearly zero. But we can factorize the interaction matrix in Eq. (43):

$$\delta\mathbf{X}_I(\omega) = -\mathbf{V}[\mathcal{G}(\omega) - \mathcal{G}(\omega)]\mathbf{X}_I(\omega) \tag{44}$$

But Eq. (44) says clearly that the error does not depend on the form of the interaction, only depends on the recording time T. Even we have supposed that the only source of error was the recording time, we do not suppose any particular behavior for the interaction. So, we have enunciated and proved a theorem:

Theorem III.

In the time-reversal problem and for left-hand material conditions, the normalized error:

$$\frac{\delta \mathbf{X}_I(\omega)}{\|\mathbf{V}\|} \tag{45}$$

is independent of the explicit form of the interaction provided the last is isotropic.

$$(\mathbf{V}^{-1} = \mathbf{V}^{t*})$$

Returning to the time representation, for the time-dependent retarded isotropic (remember that in the following expression, the indices m and n indicates components of the field and can be omitted), free Green function related to $\mathbf{K}^{(\circ)}(\omega)$, we can write explicitly.

$$G^{mn(\circ)+}(\mathbf{r},t;\mathbf{r}',t') \equiv G^{(+)}(\mathbf{r},t;\mathbf{r}',t') = \frac{\delta[t'-(t-\frac{|\mathbf{r}-\mathbf{r}'|}{c})]}{|\mathbf{r}-\mathbf{r}'|} \tag{46}$$

and for the advanced time-dependent free Green function related to $\mathbf{R}^{(\circ)}(\omega)$:

$$\mathcal{G}^{mn(\circ)-}(\mathbf{r},t;\mathbf{r}',t') \equiv \mathcal{G}^{(-)}(\mathbf{r},t;\mathbf{r}',t') = \frac{\delta[t'-(t+\frac{|\mathbf{r}-\mathbf{r}'|}{c}-T)]}{|\mathbf{r}-\mathbf{r}'|} \tag{47}$$

That is the recording time appears explicitly in the advanced Green function and we can show that its value makes possible to blend many signals on the same channel without interference. It is important to note that for resonances, the relevant Green functions are precisely the free ones and not the complete ones as we can see in Eqs. (5) and (6).

7. Information packs

In this section, we present the support and the definition of the information packs that are required for the adequate performance of the device shown in Section 6 and that by him constitute a method to improve the broadcasting efficiency. To this end, we must remember that on communication theory [9, 10] are defined the so-called ensembles of functions dependent on time. One of their properties is really a group one from the mathematical point of view and lies in that any ensemble transforms into another member of the same ensemble when we change the function at any certain amount of time. To illustrate this property, we shift by an amount t_1 the argument of all the members of the ensemble defined as follows:

$$F_\theta(t) = \sin(t+\theta) \tag{48}$$

where θ is distributed uniformly from 0 to 2π.

Then, we have:

$$F_\theta(t+t_1) = \sin(t+t_1+\theta) = \sin(t+\varphi) \tag{49}$$

where φ is distributed uniformly from 0 to 2π.

Then, each function has changed individually, but the ensemble as a whole is invariant under the transformation. Also, if we apply the operator T which gives for each member

$$S_\alpha(t) = TF_\alpha(t) \tag{50}$$

It implies that

$$S_\alpha(t + t_1) = TF_\alpha(t + t_1) \tag{51}$$

It is possible to prove that if T is an invariant operator and the input ensemble $F_\alpha(t)$ is stationary, the output ensemble $S_\alpha(t)$ is also stationary. Now, for communication purposes, the operator T, which could be a modulation process, is not invariant because of the phase carrier that gives certain time structure, but if the translations are multiples of the periods of the carrier, then the modulation will be invariant. At this stage, it is important to remember that Wiener [6] has pointed out that if a device is linear as well as invariant (in the sense of the last definition), then the Fourier analysis is the appropriate mathematical tool for dealing with the problem. Now, suppose in addition that we are interested on functions that are limited to the band from 0 to Θ cycles per second, then we have the following theorem [10]:

Let $F(t)$ contain no frequencies over Θ. Then:

$$F(t) = \sum_{-\infty}^{\infty} X_n \frac{\sin \pi(2\Theta t - n)}{\pi(2\Theta t - n)} \tag{52}$$

where,

$$X_n = F\left(\frac{n}{2W}\right) \tag{53}$$

In this expansion, $F(t)$ is represented as a sum of orthogonal (basis) functions. The coefficients X_n of the various terms can be considered as coordinates in an infinite dimensional "functions space." We will take the last theorem (Eqs. (52) and (53)) as a very suggestive rule to consider the recently obtained resonant frequencies. If we use physical arguments about the reasons of the presence of a resonance, we can be sure that channels available for broadcasting are also limited in number. Indeed, in a recent paper, we have generalized the procedure for electromagnetic scalar and vector potentials [30] and we have established that we can use either the electromagnetic field or the potentials for obtaining the resonances and also for the use of the recording time as a resource to optimize communications. And now, we can build information packs (IP) that are functions, which represent a part of the signal we want to send with the minimum loss of information. The resultant expression is:

$$F_e(t) = \sum_{-\infty}^{\infty} X_{n,e} \frac{\sin\left[\pi(2\omega_e t - n)\right]}{\pi(2\omega_e t - n)} \tag{54}$$

where,

$$X_{n,e} = F_e\left(\frac{n}{2\omega_e}\right) \tag{55}$$

Every ω_e allows us to build a decomposition like (54) but we expect that only a few terms are necessary for a well representation of $F_e(t)$. Next, we send separately each $F_e(t)$ by its own device and it is all we need for broadcasting. To receive the signal, we need a separate device for each ω_e.

A very important feature is that because of the properties of the modulation process stated in Eqs. (50) and (51), we can recover, for any arbitrary signal, the behavior under spectral representation and under separated pack representation. So we can either talk about $F_e(t)$ in Eq. (54) as the representation of some element of the basis function for the spectral representation or directly as the e component of an arbitrary signal $S(t) = TF(t)$. Now, we recall the two resonances founded in another work [3]:

$$\omega_1 = \frac{\pi}{4d} + \omega_0 \tag{56}$$

and

$$\omega_2 = \frac{3\pi}{4d} + \omega_0 \tag{57}$$

Suppose that $S(t)$ is the signal

$$S(t) = \frac{\sin\left[\pi(2\Theta t)\right]}{\pi(2\Theta t)} \tag{58}$$

Then, we have the first pack:

$$S_1(t) = \sum_{-\infty}^{\infty} X_{n,1} \frac{\sin\left[\pi(2\omega_1 t - n)\right]}{\pi(2\omega_1 t - n)} \tag{59}$$

with

$$X_{n,1} = S\left(\frac{n}{2\omega_1}\right) \tag{60}$$

And, we have the second pack

$$S_2(t) = \sum_{-\infty}^{\infty} X_{n,2} \frac{\sin\left[\pi(2\omega_2 t - n)\right]}{\pi(2\omega_2 t - n)} \tag{61}$$

with

$$X_{n,2} = S\left(\frac{n}{2\omega_2}\right) \tag{62}$$

We can see that if $\Theta = \omega_1$, the only coordinate distinct to zero is $X_{0,1} = 1$ and if $\Theta = \omega_2$, only survives the term $X_{0,2} = 1$. So, we remark self-consistency of the method.

Even VMF has a broad application on the microwave range, maybe it would be more useful to apply for larger frequencies. But even the great technological boom, there is not any device that could manipulate visible light at length as happens with microwaves. Whatever we can recall some of the basic early ideas on radio broadcasting when the option was sending information by means of modulating the wave's amplitude as appears in **Figure 3**. However, we can take our definition of information packs and put it in a modulated visible-light signal taking the enveloping of the signal we name the wrapping signal (WS) as the information that can be injected inside Eq. (54). Technically, we rewrite Eqs. (50) and (51) in the form:

$$H_\beta(t) = \Omega S_\beta(t) \tag{63}$$

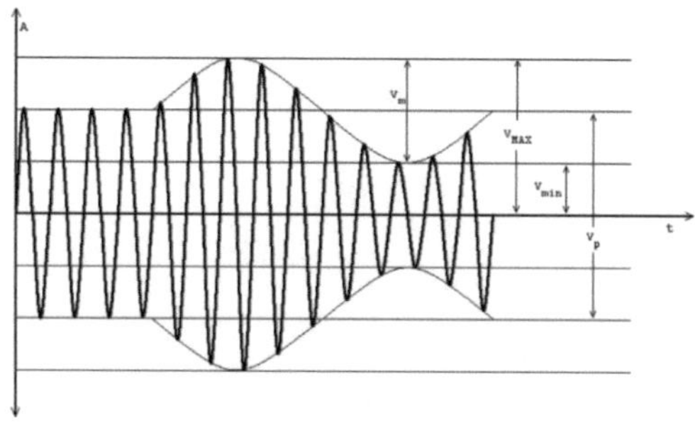

Figure 3.
The former radio broadcasting procedure: modulated amplitude. Image given by Pérez-Martinez [31].

It implies that

$$H_\beta(t + t_1) = \Omega S_\beta(t + t_1) \tag{64}$$

Now, the operator Ω is a generic operator like T but acting over the ensemble $S_\beta(t)$. Some care must be taken when reading the WS information, because the translations stated in Eqs. (63) and (64) were multiples of the periods of the carrier, and then as we said above, the modulation will be invariant. The resonant frequencies will be obtained by the same procedure.

In order to complete the methodology, we recall the concept of group velocity $c_g(t)$ and construct this inherent quotient between them and the enveloping frequency ω_g which results in the wave number κ_g, so we associate them with the resonance frequencies in a similar form as we styled with microwaves, but now these last signals come from the measured properties of the Green's function associated with the modulated signal. In this way, in Eq. (54), we put directly the WS first for a non modulated beam:

$$S_{e'}(t) = \sum_{-\infty}^{\infty} X_{n,e'} \frac{\sin\left[\pi(2\omega_{e'}t - n)\right]}{\pi(2\omega_{e'}t - n)} \tag{65}$$

in which the coefficients are given by:

$$X_{n,e'} = S_{e'}\left(\frac{n}{2\omega_{e'}}\right) \tag{66}$$

The signal $S_{e'}(t)$ in (65) can be viewed as the representation of some element of the new basis functions or as the e' component of an arbitrary amplitude-modulated signal $H_{e'}(t)$. Now, we can give an example where we use the same values for the resonances on Eqs. (56) and (57) and where we propose an arbitrary amplitude modulated or WS (for a modulated visible light beam) signal given as follows:

$$H(t) = a\cos\left(\Theta_A t + \delta\right) \tag{67}$$

In Eq. (67), $\Theta_A = \Theta_p \pm \Theta_m$ is an arbitrary frequency, and in a same manner, a and δ are preconceived constants but otherwise arbitrary.

With these preliminaries, we can build the first IP:

$$H_1(t) = \sum_{-\infty}^{\infty} X_{n,1} \frac{\sin\left[\pi(2\omega_1 t - n)\right]}{\pi(2\omega_1 t - n)} \tag{68}$$

where explicitly the coefficients are:

$$X_{n,1} = H\left(\frac{n}{2\omega_1}\right) \tag{69}$$

And taking expression (67)

$$X_{n,1} = a\cos\left[\Theta_A\left(\frac{n}{2\omega_1}\right) + \delta\right] \tag{70}$$

In a similar manner, the second IP will be:

$$H_2(t) = \sum_{-\infty}^{\infty} X_{n,2}\frac{\sin\left[\pi(2\omega_2 t - n)\right]}{\pi(2\omega_2 t - n)} \tag{71}$$

in which

$$X_{n,2} = H\left(\frac{n}{2\omega_2}\right) \tag{72}$$

Also, by taking Eq. (67):

$$X_{n,2} = a\cos\left[\Theta_A\left(\frac{n}{2\omega_2}\right)\delta\right] \tag{73}$$

As we said above, the resonances must come also for the WS. By this procedure, we have enlarged the scope of the formalism we named vector-matrix or VMF [1–3].

In order to complete our example, we put explicit values of the resonances for the two visible light IP:

$$H_1(t) = \sum_{-\infty}^{\infty} X_{n,1}\frac{\sin\left[\pi\left(2\left[\frac{\pi}{4d} + \omega_0\right]t - n\right)\right]}{\pi\left(2\left[\frac{\pi}{4d} + \omega_0\right]t - n\right)} \tag{74}$$

And explicitly

$$X_{n,1} = a\cos\left[\Theta_A\left(\frac{n}{2\left[\frac{\pi}{4d} + \omega_0\right]}\right) + \delta\right] \tag{75}$$

For the second IP

$$H_2(t) = \sum_{-\infty}^{\infty} X_{n,2}\frac{\sin\left[\pi\left(2\left[\frac{3\pi}{4d} + \omega_0\right]t - n\right)\right]}{\pi\left(2\left[\frac{3\pi}{4d} + \omega_0\right]t - n\right)} \tag{76}$$

in which

$$X_{n,2} = a\cos\left[\Theta_A\left(\frac{n}{2\left[\frac{3\pi}{4d} + \omega_0\right]}\right) + \delta\right] \tag{77}$$

8. Conclusions

In Eqs. (25), (29), (30), (34)–(40), we have shown that it is possible to use an operator language and the properties of the Green function to define the capacity of a channel, the loss of information, and finally, the error in the time-reversal process. Therefore, we can use our results to describe the behavior of LHM interacting with electromagnetic field whether forward or backward in time. Thanks to our interpretation of a resonance in the broadcasting problem with the left-hand material conditions, and the application of the model PSM, we make up a broadcasting system that has the power for distinguishes between signals according to their recording time, and allows to superpose signals in the same frequency range having different recording times with the minor loss because of resonance technology; to this end, we have presented a detailed support and definition of the information packs (IP) and the possibility of application for visible light. In addition, we have enunciated and proved a theorem (theorem III) that establishes: for the TRT and LHM, the normalized error is independent of the particular behavior of the interaction. Summarizing, we give a complete recipe for optimizing communications efficiency.

Author details

Juan Manuel Velazquez Arcos*, Ricardo Teodoro Paez Hernandez, Tomas David Navarrete Gonzalez and Jaime Granados Samaniego
Universidad Autónoma Metropolitana, Mexico City, Mexico

*Address all correspondence to: jmva@correo.azc.uam.mx

References

[1] Velázquez-Arcos JM, Granados-Samaniego J. Wave propagation under confinement break. IOSR Journal of Electronics and Communication Engineering (IOSR-JECE). 2016;**11**(2. Ver. I):42-48. e-ISSN: 2278-2834, p-ISSN: 2278-8735. Available from: www.iosrjournals.org

[2] Velázquez-Arcos JM, Granados-Samaniego J, Vargas CA. Electromagnetic scalar and vector potentials on the problem of left-hand materials conditions. Electromagnetic in Advanced Applications. 2016;**1**:1. edited by IEEE

[3] Granados-Samaniego J, Velázquez-Arcos JM, Vargas CA, Tavera Romero F, Hernández López RT. Resonant technology and electromagnetic packaging. In: International Conference on Electromagnetic Advanced Applications 2015; Torino, Italia; 2015

[4] Velázquez-Arcos JM, Granados-Samaniego J, Vargas CA. Fredholm and Maxwell equations in the confinement of electromagnetic field. In: International Conference on Electromagnetic Advanced Applications 2015; Torino, Italia; 2015

[5] Velázquez-Arcos JM, Granados-Samaniego J, Vargas CA. Resonance regime on a plasma sandwich simulates a left-hand material broadcasting condition. In: Electromagnetics in Advanced Applications (ICEAA), 2014 International Conference; 2014. pp. 137-140. DOI: 10.1109/ICEAA.2014.6903842

[6] Velázquez-Arcos JM. Fredholm's alternative breaks the confinement of electromagnetic waves. AIP Advances. 2013;**3**:092114. DOI: 10.1063/1.4821336

[7] Velázquez-Arcos JM, Granados-Samaniego J, Vargas CA. The confinement of electromagnetic waves and Fredholm's alternative, Electromagnetics in Advanced Applications (ICEAA). In: 2013 International Conference; 2013. pp. 411-414. DOI: 10.1109/ICEAA.2013.6632268

[8] Kong X-k, Liu S-b, Zhang H-f, Bian B-r, Li H-m, et al. Evanescent wave decomposition in a novel resonator comprising unmagnetized and magnetized plasma layers. Physics of Plasmas. 2013;**20**:043515. DOI: 10.1063/1.4802807

[9] de Rosny J, Fink M. Overcoming the diffraction limit in wave physics using a time reversal Mirror and a novel acoustic sink. Physical Review Letters. 2002;**89**(12):124301

[10] Lerosey G, de Rosny J, Tourin A, Fink M. Focusing beyond the diffraction limit with far-field time reversal. Science. 2007;**315**:1120-1122

[11] Velázquez-Arcos JM. Fredholm's equations for subwavelength focusing. Journal of Mathematical Physics. 2012; **53**(10):103520. DOI: 10.1063/1.4759502

[12] Velázquez-Arcos JM, Pérez-Martínez F, Rivera-Salamanca CA, Granados-Samaniego J. On the application of a recently discovered electromagnetic resonances to communication systems. IJETAE. 2013; **3**(1):466-471. Website: www.ijetae.com. ISSN: 2250-2459

[13] Velázquez-Arcos JM, Granados-Samaniego J, Vargas CA. Communication theory and resonances on electromagnetic systems, Electromagnetics in Advanced Applications (ICEAA). In: 2012 International Conference; 2012. (IEEE Cape Town). pp. 392-395, DOI: 10.1109/ICEAA.2012.6328657

[14] Velázquez-Arcos JM, Granados-Samaniego J. Recent technologies and

recording time detection on time reversal signals. International Journal on Recent and Innovation Trends in Computing and Communication. 2014;**2**(11): 3447-3950

[15] Grbic A, Eleftheriades GV. Negative refraction, growing evanescent waves, and sub-diffraction imaging in loaded transmission-line Metamaterials. IEEE Transactions on Microwave Theory and Techniques. 2003;**51**(12):2297-20305

[16] Xu H-X, Wang G-M, Lv Y-Y, Qi M-Q, Gao X, Ge S. Multyfrequency monopole antennas by loading metamaterial transmission lines with dual-shunt branch circuit. Progress In Electromagnetics Research. 2013;**137**: 703-725

[17] Kato H, Inoue M. Reflection-mode operation of one-dimensional magnetophotonic crystals for use in film-based magneto-optical isolator devices. Journal of Applied Physics. 2002;**91**:7017-7019

[18] Kato H, Matsushita T, Takayama A, Egawa M, Nishimura K, Inoue M. Theoretical analysis of optical and magneto-optical properties of one-dimensional magnetophotonic crystals. Journal of Applied Physics. 2003;**93**: 3906

[19] Hernández-Bautista F, Vargas CA, Velázquez-Arcos JM. Negative refractive index in split ring resonators. Revista Mexicana de Fisica. 2013;**59**(1): 139-144. ISSN: 0035-00IX

[20] Velázquez-Arcos JM. Nanotechnology can be helped by a new Technology for Electromagnetic Waves. Nanoscience and Nanotechnology. 2012; **2**(5):139-143. DOI: 10.5923/j.nn.20120205.02

[21] de la Madrid R. The decay widths, the decay constants, and the branching fractions of a resonant state. Nuclear Physics A. 2015;**940**:297-310

[22] Velázquez-Arcos JM, Vargas CA, Fernández-Chapou JL, Salas-Brito AL. On computing the trace of the kernel of the homogeneous Fredholm's equation. Journal of Mathematical Physics. 2008; **49**:103508. DOI: 10.1063/1.3003062

[23] de la Madrid R. The rigged Hilbert space approach to the Gamow states. Journal of Mathematical Physics. 2012; **53**(10):102113. DOI: 10.1063/1.4758925

[24] Velázquez-Arcos JM. Fredholm's equation and Fourier transform on discrete electromagnetic systems. IJRRAS. 2012;**11**(3):456-469. Available from: www.arpapress.com/Volumes/Vol11Issue3/IJRRAS_11_3_11.pdf

[25] Shannon CE. A mathematical theory of communication. The Bell System Technical Journal. 1948;**27**:379-423, 623-656

[26] Nyquist H. Certain factors affecting telegraph speed. Bell System Technical Journal. 1924:324. Certain Topics in Telegraph Transmission Theory, A.I.E.E. Trans., vol. 47, April 1928, pp. 617

[27] Hartley RVL. The Interpolation, Extrapolation and Smoothing of Stationary Time Series, NDRC Report. New York-London: Wiley; 1949

[28] Wiener N. The Ergodic theorem. Duke Mathematical Journal. 1939;**5**:1-18

[29] Hopf E. On causality statistics and probability. Journal of Mathematical Physics. 1934;**13**(1):51-102

[30] Velázquez-Arcos JM, Granados-Samaniego J, Vargas CA. Resonances and different recording times on electromagnetic scalar and vector potentials. In: 2017 Progress in Electromagnetics Research Symposium-Fall (PIERS-Fall), Singapore; 2017. pp. 229-235

[31] Pérez-Martinez F. Fundamentos de Sistemas de Televisión. México, D.F.: Universidad Autónoma Metropolitana; 2009

Chapter 3

Android Application Security Scanning Process

Iman Almomani and Mamdouh Alenezi

Abstract

This chapter presents the security scanning process for Android applications. The aim is to guide researchers and developers to the core phases/steps required to analyze Android applications, check their trustworthiness, and protect Android users and their devices from being victims to different malware attacks. The scanning process is comprehensive, explaining the main phases and how they are conducted including (a) the download of the apps themselves; (b) Android application package (APK) reverse engineering; (c) app feature extraction, considering both static and dynamic analysis; (d) dataset creation and/or utilization; and (e) data analysis and data mining that result in producing detection systems, classification systems, and ranking systems. Furthermore, this chapter highlights the app features, evaluation metrics, mechanisms and tools, and datasets that are frequently used during the app's security scanning process.

Keywords: Android, application, scanning, security, malware

1. Introduction

This section introduces the Android operation system and its applications. Moreover, it defines Android malware and shares its recent statistics. Android permissions and security model are also presented. This section ends with discussing the security scanning framework for Android applications.

1.1 Android and application definition

Android is one of the most popular operating systems that provide open-source development environment based on Linux. It allows the development for mobile, tablets, smartwatches, and smart TVs. Android was established by Open Handset Alliance that started working in 2003, while Google released its first Software Development Kit (SDK) in 2007, but the first commercial version was released in September 2008 called as Android 1.0 [1] with the first device executed being HTC Dream. The sale of the Android phone was increased from 75% in 2013 [2] to 88% in 2018 [3]. **Table 1** lists the sales of smartphones from 2011 to 2018 which show a clear capture of the market over the years. This market penetration reveals the successful implementation of features as well as cheap price.

The Android system is composed of five important layers:

- **Applications** refer to the software stack of native as well as user-based applications.

IntechOpen

- **Android runtime** allows the application to run on mobile devices by converting the Android code into DEX format or byte code. The conversion of DEX code into device-related code is done before compilation, and this kind of technique is referred to as ahead of time (AOT).

- **Application framework** manages and runs the applications using the services such as activity manager, content providers, telephony manager, package manager, location manager, etc.

- **Android libraries** are a set of Java-based development application programming interfaces (APIs) that can help in performing general purpose tasks, as well as location-based and string handling.

- **Android kernel** is based on the Linux 2.6 kernel and is used to provide abstraction between device hardware and other software layers [4, 5].

The efforts for making each of the component secure have been made. However, still there are issues due to open-source development, and every vendor and company following their own standards has led to serious security issues [6].

The Android application contains four types of components shown in **Figure 1** [7]:

- **Activities:** each activity represents a single screen with a user interface.

- **Services:** a service operates in the background to execute long-running operations. Services could be initiated by other components like activity or broadcast receiver.

- **Content providers:** to share data between different applications.

- **Broadcast receivers:** to listen for specific system-wide broadcast announcements and react to them.

Android applications are written in Java programming language and distributed as .apk files. Android application package (APK) file is a ZIP compressed file that includes the following files:

- **AndroidManifest.xml file**: it describes the application's capabilities and informs the OS about the other components of the application. It identifies the needed hardware and software features such as the camera, in addition to, the minimum API level required by the application. The permissions requested by the app and the permissions required to access the application's interfaces/data are defined in its manifest file.

- **Dalvik executable or classes.dex file**: the Java classes and methods defined in the application code are grouped into one single file (classes.dex).

- **.xml files**: which are used to define the user interface of the application.

- **Resources**: the external resources that are associated with the application (e.g., images).

Android applications run in a virtual environment to improve security. However, they can be downloaded from any source whether trusted or not. After an application is initiated, it grants its own virtual environment, so the code will be isolated

Year	Android share	iOS shares	Other OS shares
2011	46.66	18.87	34.45
2012	66.34	19.11	14.53
2013	78.50	15.54	5.94
2014	80.70	15.37	3.91
2015	81.60	15.88	2.50
2016	84.79	14.44	0.75
2017	85.91	13.98	0.09
2018	88	11.75	0.03

Table 1.
The detail of the Android phone compared to iOS and other smartphone sales shares from 2011 to 2018 retrieved from Statista.com [3].

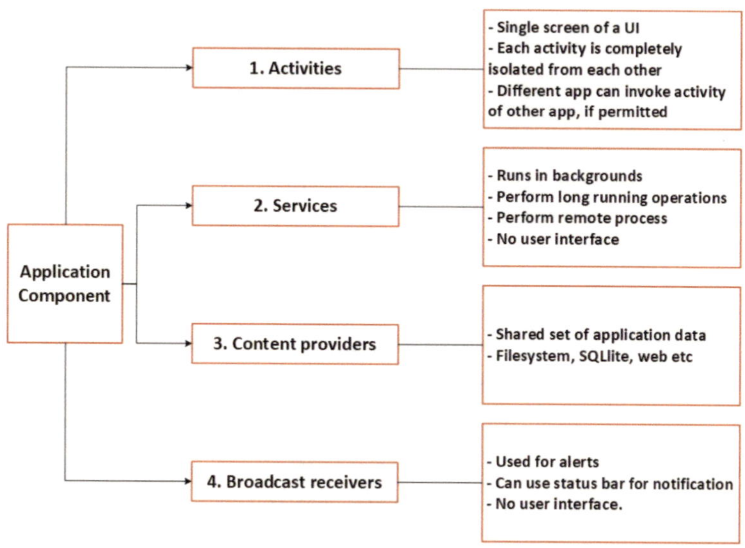

Figure 1.
The Android application components.

from other apps. Although the applications are isolated, still they can interact with the system and other applications through APIs. Meanwhile, Android assigns Linux user ID for each application.

API stands for application programming interface that refers to the set of tools providing interfaces for communication between different software components. APIs are used to access data and key features within Android devices. API framework consists of a set of API packages that include specific classes and methods. Additionally, it contains a set of XML elements and attributes for declaring a manifest file and accessing resources besides permissions and intents. Looking into API component calls in the executable file may allow exploring the behavior of an app and reporting its capabilities. However, in many cases, the attackers hide the API calls using cryptography, reflection, or dynamic code loading techniques to increase the difficulty of analysis.

1.2 Malware definition and statistics

Presently, mobile device apps are distributed through online marketplaces such as Google Play Store. Such marketplaces are considered hubs to allow developers to publish their apps and distribute them as well. Today, there are more than 2.5 million applications available in the Google Play Store [8].

When downloading apps from unofficial markets, the user is usually at risk because there is no centralized control like official markets. As more users shift to Android devices, cybercriminals are also turning to Android to inflate their gain. However, many Android apps turn out to be malicious. The number of malicious software (malware) samples in the Android market has surged to an alarming number reaching over 5.49 million by the end of 2018 [9, 10].

A recent report from F-Secure [11] showed that over 99% of all malware programs that target mobile devices are designed for Android devices. Another report from the security firm G DATA shows that a new instance of Android malware pops up nearly every 10 seconds. Another report from AV-TEST [12] states very clearly that anyone seeking to make money by attacking mobile devices will choose Android devices as targets.

Malware is an umbrella term used to stand for an assortment of types of hostile or intrusive software, including viruses, worms, Trojan horses, ransomware, spyware, adware, and different malicious programs [13].

Ransomware is considered one of the most threating malwares nowadays. There are two types of ransomware: crypto ransomware and lock screen ransomware. The crypto ransomware encrypts the information, while the locker ransomware hinders users from gaining access to their data by locking the device's screen. For both types, the attack demands a payment (ransom) to recover the files or access to the device. It is worth mentioning that paying the ransom money does not guarantee that the files will be back or that the ransomware will be removed from the device [14, 15].

According to Kaspersky, ransomware has taken place in most of the majority of notorious security attacks for the past decade. Also, 116.5 million attacks were noted in 2018, compared to 66.4 million in 2017, an increase of twofold in just 1 year [16].

Malicious apps, in general, are distributed mostly through phishing, drive-by attacks, and app stores. Phishing messages might comprise links to malicious apps and are sent over SMS or WhatsApp. Drive-by attacks are carried out by Web page exploits. When the user has a vulnerable browser, the exploit is able to execute a code. To infect users through app stores, malwares are submitted to them hiding as some legitimate app. In fact, in some cases some popular apps are modified to include malicious actions while keeping the app's main functionalities [17].

Therefore, a reliable tool is needed to test the trustworthiness of these apps before being installed. App risk scoring or rating should be empirically calculated according to different risk scoring techniques. The visualization of these risks should be easy enough for a normal user to recognize the risk associated with a specific app.

1.3 Android permissions and security model

Android platform is very popular due to its available and comprehensive API framework [18]. Android API offers the developers of mobile apps the ability to gain access to hardware information, accessing user's data, knowing phone state, changing phone settings, etc. The developers are impacted by the permission model while developing mobile applications. To develop a mobile app, the engineers are required to determine, for each API functionality, what permissions are needed and how they are correctly activated. Android asks the developers to list publicly what permissions are used by the app; however, there are no mechanisms to know the exact purpose of such permissions and what kind of sensitive data they could use.

Android permissions mainly fall under four categories [19]:

- **Normal:** minimal risk permission is assigned automatically by the system and does not require an explicit declaration.

- **Dangerous:** the permission to private data, system process, and other hardware is referred to as dangerous and should be assigned explicitly at the time of installation or usage of the application.

- **Signature:** the applications get the same ID and the same access rights if the two application certificates are the same.

- **SignatureOrSystem:** the applications that are signed with the same certificate will get the same permission as the base system automatically.

Take the camera permission as an example; it belongs to the dangerous category. These permissions also ensure the safety of the system by keeping the user aware of what he is trying to do and what permissions have been requested. The issue with Android permission is that they are coarse-grained and violate the principle of least privileges (PoLP) that ensure that the only required thing is permitted. In contrast, Android allows overall permission about most of the features such as phone contact permission can allow checking phone state and other details.

The Android permission system obliges app's developers to state which security critical resources are needed. At runtime, the access requests are controlled by the permission checker component in order to secure the critical resources and operations.

In general, the security policy for the phones is delegated to their users. The lists of permissions will be shown to the users where they can accept or reject. It is essential and challenging to make sure that these apps appropriately deal with great value sensitive data [18].

Since Android 6.0, dangerous permissions are now asked explicitly to the user when requested the first time and then granted automatically. Android 6.0 changed some areas with regard to permissions. Two major changes were introduced. (1) Apps targeting SDK 23 (Software Development Kit) or higher can request permissions at run-time. (2) If an app requests a dangerous permission, with another permission from the same group that has been already granted, the system immediately grants it without any further interaction with the user.

The Android permission system received several criticisms [20]. The system is considered to be too coarse-grained since the user has to choose whether to accept all of the permissions declared by an app or to refuse to install the app. Users are usually not sure to determine if an app can be trusted or not. Actually, how Android is showing the required permissions is not very user-friendly and quite difficult to understand the risks associated with these permissions.

Android apps are allowed to define new custom permissions on Android. These permissions are used to protect an app's own resources from others. To define new custom permission, a permission name is needed, optionally including a permission group and a description regarding the permission purpose. Sixty-five percent of Google Play Store apps define their custom permissions, whereas 70% of these apps request them for their operation [21].

Mobile app history has shown that the users' privacy and security must be protected against benign applications not to mention malware ones. Actually, lots of widely used apps have been reported as requiring too many permissions or leaking user information to their servers intentionally [22].

Android uses both discretionary access control (DAC) and mandatory access control (MAC) to form a multilayer security model [23]. The model implements a kernel-level application sandbox that uses Linux user identifiers (UIDs) and UNIX-style file permissions. Since version 4.3, Security-Enhanced Linux (SELinux) was introduced, and from version 4.4 it started being deployed in enforcing mode.

Android security has seen other improvements as well. In version 5.0, Google introduced smart lock, which allows users to unlock the phone using a trusted device, such as Bluetooth/NFC beacon, smartwatch, or facial recognition. In version 6.0, they introduced a fingerprint API. All these features are an extra step to make security easier for the average user. However, Android's security model is still based on a set of coarse-grained permissions.

Android builds its security basis on multiuser capabilities of Linux by assigning a unique ID to each application that will manage its own processes [24]. The runtime manager runs the applications in its sandbox that provides security as it does not allow:

- Inter-process communication
- Data access to other processes
- Hardware access such as camera, GPS, or network
- Access to local data of the phone such as media libraries and contacts

As a contrast to other OS platforms, the sandbox facility is provided by runtime manager for direct access to resources and hardware, while other operating systems provide sandboxing based on their kernel. This is based on features such as all kinds of requests outside the applications are by default denied and have to be permitted explicitly. When an application is installed, the permissions are allocated in addition to a unique, permanent identification (ID) that is also assigned to this app. This application ID is used to enforce the permissions for application, processes, and file system [25–27].

The files in the application are always private unless they are explicitly set to be shared using two modes, (1) readable and (2) writable. In order for two applications to share other's files, then the application ID must be the same for both applications, as well as the user ID. Additionally the public key infrastructure (PKI) certificate value must be shared to be considered as one application [27]. The paranoid network security mechanism is used to protect the network access by keeping all kinds of network access in separate groups such as WiFi, the Internet, and Bluetooth. Thus, if the application or process gets the permission to access a Bluetooth, then its application ID will be added to the group access list for Bluetooth and similarly for others. Consequently, one application can be assigned to one or more access groups [28].

Before any application distribution, Google that manages the main play store requires to sign the application using developers' personal certificate to make sure that the distributed copy is done through the right developer and no modification can be made to the application. If the two application matches the same certificate, Android will assign the same application ID to both applications and will access to private files for each application [27, 28].

Relying only on the current Android security model and permission levels to secure Android app is inefficient. Other more comprehensive security systems need to be considered and implemented to ensure efficient detection of malware apps. Consequently, the following sections present a reference model for Android application's security scanning process.

1.4 Android application scanning framework

A reference model for Android scanning process is shown in **Figure 2**. This model provides the core steps/phases vital to analyze Android apps and malware detection. The following sections highlight each one of these phases, starting from allocating the source of Android apps, downloading mechanisms, app's source code generation process, app's features extraction, applying static and dynamic analysis, generating datasets, detecting and classifying the app into benign or malware, and ranking its risk if it is detected as a malware app. Moreover, the mostly used mechanisms and tools utilized by researchers and developers at each process's phase are also presented.

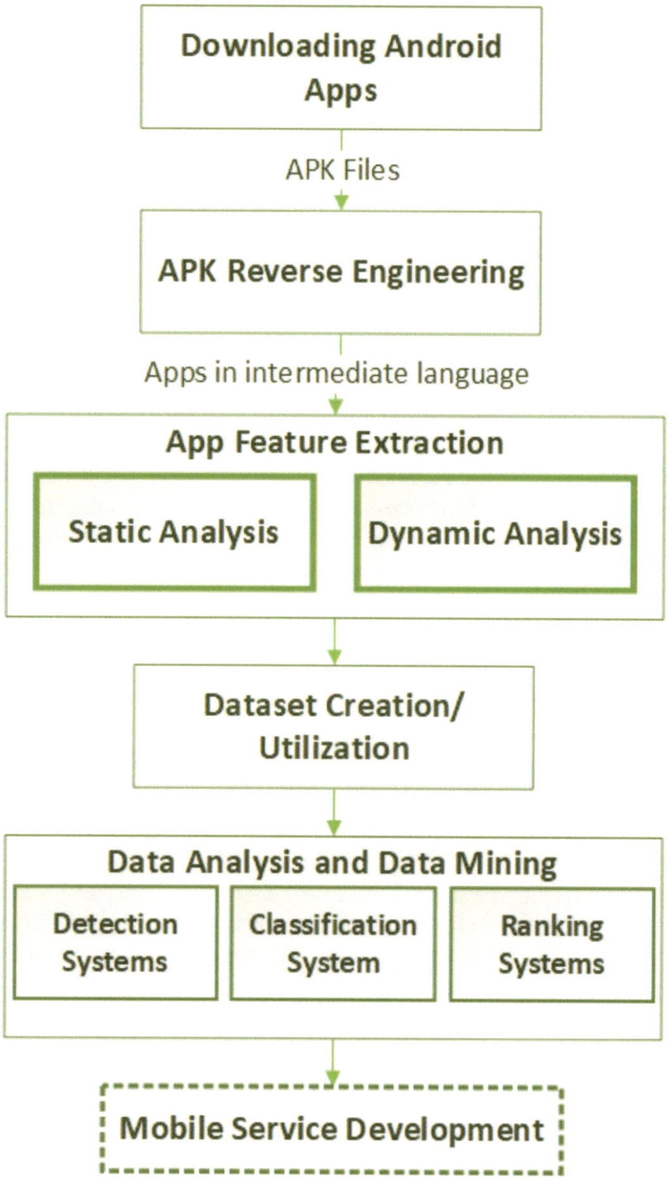

Figure 2.
Android application security scanning model [29].

2. Android application

The app's source and how they are downloaded are presented in this section, in addition to the source code generation for Android applications.

2.1 App source and download

Android application collection process includes gathering APK files from different Android marketplaces. The main application sources include Google Play, Anzhi, and AppChina. For every potential free app, the crawler script must ensure that the app has not been downloaded before and then calculate the app's hash using the SHA256 algorithm [30]. Once the app is downloaded, it can be archived for future use. Chrome APK Downloader, a desktop version of APK downloader tool, can be used to download the APK files of the free Android applications into desktop from Google Play marketplace [31]. For the paid applications, the Raccoon APK Downloader can be used to download APK files from Google Play Store [32].

2.2 Source code generation

After downloading the apps, they need to be analyzed. In order to do that, APK reverse engineering process is required to decompile, rebuild, and convert the Android executable code (.apk file) into an intermediate language such as Smali, Jimple, and Jasmin [33]. The aim of reverse engineering is to retrieve the source file from the executable files in order to apply program analysis. Unzipping the APK files generates .dex files. By reassembling the dex files using an APK reverse engineering tool, the Java files can be retrieved. Three of the most popular tools that have been used in Android APK reverse engineering are Apktool, Dex2jar, and Soot. A comparison of Android reverse engineering tools was conducted in [33]. The results showed that Apktool which uses Smali reassembled 97% of the original code, whereas Soot which uses Jimple and Dex2jar which uses Jasmin preserve 73% and 69% of the app's original code, respectively.

3. Android application analysis

The process of analyzing Android apps to detect different types of malwares and the result of such analysis in terms of datasets are illustrated in this section.

3.1 Feature extraction

Once the app's source code is retrieved, the feature extraction process starts. The features that are usually extracted depend on the type of malware and the analysis mode whether static or dynamic. This will be explained in the following two subsections. **Table 2** lists the most commonly used static and dynamic features [34].

3.2 Static analysis

The static analysis aims to check the existence of malware by disassembling the source code without executing the application. Tools which perform static analysis are mainly categorized into three approaches as shown in **Figure 3**: (1) signature-based detection, (2) permission-based detection, and (3) Dalvik Bytecode detection. There is some limitation which is related to each static detection approach. The

Static features	Dynamic features
Permission	Network, SMS, power usage, CPU, process info, native and Dalvik memory
API calls	Data packets being sent, IP address, no. of active communications, system calls
String extracted	Process ID, system calls collected by strace, returned values, times between consecutive calls
Native commands	Network traffic, destination IP address
XML elements	System calls collected by strace, logs of system activities
Meta data	Data collected by logger, Internet traffic, battery percentage, temperature collected every minute
Opcodes from .dex file	
Task intents	

Table 2.
Most commonly used features in static and dynamic analyses [34].

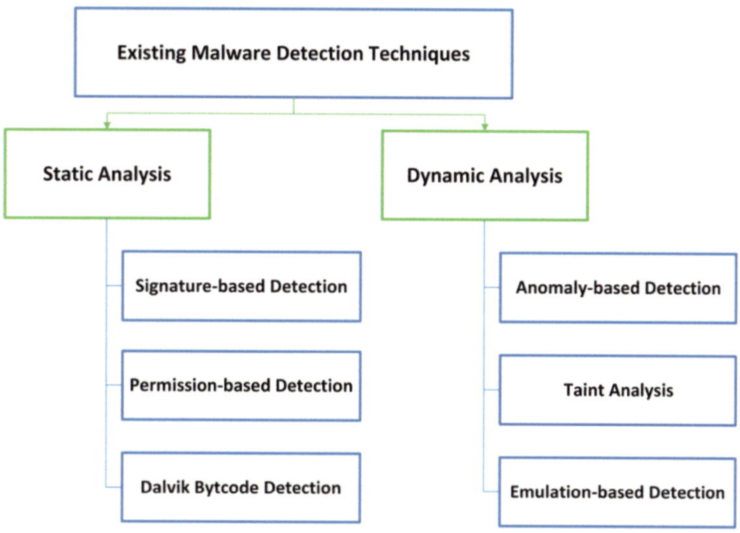

Figure 3.
Existing malware detection techniques.

signature-based detection which relies on stored signatures for known malwares does not have the ability to detect unidentified malware signatures [35]. In the permission-based detection, the benign app could be considered, incorrectly, as a malware due to the minor variation of the requested permissions from the original and the malware application [30]. Finally, the Dalvik Bytecode detection, which assists in evaluating the applications actions, consumes more resources [36]. Several tools were discussed and analyzed in [35–41] such as FlowDroid [39], PScout [40], and ApkAnalyser [41]. Each of the aforementioned tools focuses on one or more features. The most generally extracted static features are the permissions [18], API calls [42], and source code metrics [43].

Years ago, different rival static approaches have been proposed like TaintDroid [43], DroidRanger [45], and RiskRanker [46] to detect malicious malware features. But all of them rely on manually crafted detection patterns which may not be able to detect new malware and come with significant device performance cost [47].

Authors in [48] proposed DREBIN as the first approach which provides detection of Android malicious code directly on the mobile device. They used static analysis in a machine learning system to distinguish malware from trusted applications. They considered linear support vector machines for classification. This approach, however, cannot detect runtime loaded and obfuscated malicious applications because it relies on static analysis [47].

Yang [49] developed a prototype (AppContext) that detects malicious apps using static analysis. They mined 633 benign apps from Google Play and 202 malware apps from various malware datasets. AppContext identifies applications using machine learning based on the contexts that trigger security-sensitive behaviors (e.g., the conditions and events that cause the security-sensitive behaviors to occur). But this approach can be evaded by dynamic code loading and consumes huge human efforts in labeling each security-sensitive behavior [50].

Akhuseyinoglu and Akhuseyinoglu [51] have proposed an automated feature-based static malware detection system called AntiWare for Android devices. It is automated since it engages the machine learning method for detecting malicious applications by using the extracted apps' features. They took into consideration the requested permissions and Google market data, including developer name, download time, and user ratings from Google Play as a feature set. AntiWare is designed to predict the rank of an application inquired by the user as malicious or benign and then report the results to the user. The main disadvantages are primarily depending on market data on Google Play and the requested permissions. The market data is not reliable since a lot of applications are invented by different new developers every second. Additionally, the permissions by its own are not sufficient to assess the malicious behavior of an application.

3.3 Dynamic analysis

In dynamic analysis, the application actions are dynamically analyzed and monitored during the execution time. The unexecuted code might be missed by this approach, but it can effectively detect the malware behaviors which are not detectable by the static analysis. Since this approach occurs during runtime, it can be performed in a controlled environment to avoid damaging the device [52].

Android dynamic malware analysis detection techniques (see **Figure 3**) can be classified into [53, 54]:

- **Anomaly detection:** the anomaly-based detection has the ability to identify suspicious behaviors to indicate the presence of malware. A drawback for this technique is that it can sometimes flag a benign application as malware because it displayed similar behaviors of malware.

- **Taint analysis:** it is an efficient technique that checks and monitors sensitive information; however, a limitation is that the performance becomes very slow rendering it useless to be applied in real time.

- **Emulation-based detection:** it is a detection technique, where it scans the application behavior by simulating the conditions of its execution environment to determine if the application is a benign or malware application from the behavior. Similar to this technique is sandbox-based detection, but the main difference originates from the details of designing each approach. A major drawback for this approach is that it requires more resources.

Tam [55] applied dynamic analysis method and machine language to detect malware. They capture real-time system calls performed by the application as key information to discriminate between ransomware, malware, and trusted files and called it CopperDroid. CopperDroid runs the Android application in the sandbox and records all system calls, in particular inter-process communications (IPC) and remote procedure call (RPC) interactions which are essential to understanding an application maliciousness behavior. However, some types of malware can detect the virtual environment and act differently (as a benign) which gives false positives.

Recent research [56] dynamically classifies Android applications to malicious or benign in the first launching of the app. The classification is applied based on the frequency of system calls as an indicator of suspicious behavior. They have built a syscall-capture system to capture and analyze the behavior of system call traces made by each application during their runtime. They have achieved an accuracy level of 85% and 88% using the decision tree algorithm and the random forest algorithm, respectively.

Also, Wang [57] proposed a dynamic analysis to analyze an application on the fly to detect malicious behavior. They developed a prototype called Droid-AntiRM to identify malware applications that employ anti-analysis techniques. The prototype identifies the condition statements in applications that could trigger the malicious acts of malware, which are unable to be recognized by static analysis. However, their prototype cannot handle dynamic code loading, encryption, or other various obfuscation techniques.

Many tools have been developed based on the dynamic perspective such as TaintDroid [44], Droidbox [58], and MobSF [59]. Additionally, some tools are considering both static and dynamic analysis in their solutions such as VirusTotal tool [60].

3.4 Ransomware detection

Unfortunately, there were very few researches studying ransomware where the malicious app blocks access to the Android device or/and its data. In [61] the authors presented a tool called Cryptolock that focuses on detecting ransomware by tracking the changes in real-time user data. They have implemented the tool on Windows platform. However, Cryptolock may send a false-positive alert because it cannot differentiate whether the user or the ransomware is encrypting a set of files [62]. They focus on changes on user's data rather than trying to discover ransomware by investigating its execution (e.g., API call monitoring and access permissions).

HelDroid tool [63] was developed to analyze Android ransomware and to detect both crypto and locking ransomwares. The tool includes a text classifier that uses natural language processing (NLP) features, a lightweight Smali emulation technique to detect the locking scheme, and the application of taint tracking for detecting file-encrypting flows. The primary disadvantage of this approach is that it highly depends on a text classifier as it assumes the availability of text. Also, it cannot be applied to some languages that have no specific phase structure like Chinese, Korean, and Japanese. This approach can be easily avoided by ransomware by applying techniques such as encryption and code obfuscation [63]. Moreover, like whatever machine learning approach, HelDroid trains the classifier in order to label an app as a ransomware. The detection capability of the model depends on the training dataset [64–66].

Another work in literature exploring the ransomware detection in Android mobiles was presented in [67]. The authors presented R-PackDroid as a static analysis approach that classifies Android applications into ransomware, malware, or benign using random forest classifier. The classification employed was based on information taken from the system API packages. An advantage over the previous approach (HelDroid) is its ability to detect ransomware regardless of the application language. Also, it flags the applications that were recognized as ransomware with very low confidence by the VirusTotal service. However, R-PackDroid cannot analyze applications with a feature code that is dynamically loaded at runtime or classes that are fully encrypted because it relies on static analysis.

Likewise, Mercaldo [68] focused on ransomware detection specifically in Android. They tested a dataset composed of 2477 samples with real-world ransomware and benign applications. The main issue of this approach is that it is manual and requires a lot of effort to analyze and build logic rules used for the classification [69].

Another automated detection approach was introduced in [70] to analyze and penetrate the malicious ransomware. They have introduced some features of static and dynamic analysis of malware. In static analysis, malicious features can be discovered with permissions, API calls, and APK structure, while malicious features in the dynamic analysis may include access to sensitive data or sensitive paths, access to the HTTP server, and user charge without notification and bypass permissions. The aim was to produce a better performance apparatus that supports ransomware detection in Android mobiles which they have designed but did not implement. The authors analyzed one malware and listed the steps of APK analysis as a concept but did not implement the proposed design. Therefore, there are no results that can prove the effectiveness of their approach.

In [71], the authors experimentally presented a new framework called DNADroid which is a hybrid of static and dynamic techniques. This framework employs a static analysis approach to classify apps into suspicious, malware, or trusted. Only suspicious classified applications are then inspected by dynamic analysis to determine if it is ransomware or not. The main weakness is that dynamic analysis is only applied to suspicious applications leaving the possibility of having malware not successfully recognized by static analysis.

3.5 Dataset creation and utilization

Datasets are mainly in two types. The first type is the Android application datasets. These include both benign apps and malicious apps. For the benign apps, the majority of researchers are collecting them from the app stores like Google Play Store [30, 37, 60]. For malicious apps, it depends on malicious behavior under study. For example, for malware Android apps, VirusTotal was one of the main sources for many researchers [38, 60]. For ransomware apps, HelDroid project [63] and RansomProper project [38] were also used.

The second type is the datasets generated after extracting the app's features. The researchers can either use existing constructed datasets considering the features under study or build new ones by screening the apps and extracting their features. The main concerns regarding the use of existed datasets are (1) absence of up-to-date apps and operating system version (2) including many duplicate samples (3) and not being accessible. These reasons could motivate researchers to build their own up-to-date and labeled datasets.

4. Android malware application detection and ranking

Many previous works have considered the problem of ranking Android apps and classify them to either malware or benign apps. The majority of these solutions have relied mainly only on the permission model and what types of permissions are requested/used by the application. They used different ways and depth of analysis in this regard.

The work presented in [72] studied the permission occurrences in the market apps and the malware apps. Also, the authors analyzed the rules (a combination of permissions) defined in Kirin [73] in order to calculate the risk signals and to reduce the warning rates. Gates et al. in [74] have compared work presented in [72, 73], naïve-based algorithms and two proposed methods for risk scoring. These methods consider the rarity of permissions as the primary indicator that contributes to raising a warning. The performance comparison was in terms of the detection rate.

The authors in [75] used similar hypotheses of listing the permissions in each app and count occurrences of permissions in similar apps (a game category in their case) excluding the user-defined permissions that are not affecting the privacy. In their solution, they gave the user the choice to turn off the permission(s). In [76], the authors used the combination of features (permissions) to compare the clean app values with the malware values to come up with thresholds that will be used to classify new Android apps. Within the same context, the idea presented in [77] was to construct a standard permission vector model for each category, which can be used as a baseline to measure and assess the risk of applications within the same category. For each downloaded app, the permission vector will be extracted and compared with the standard one; the amount of deviation from the baseline will calculate the app's risk.

While discussing the approaches in the existing risk scoring systems and their main dependency on the Android permissions, it is worth asking how many of them have considered the involvement of the user with the scoring results and, if they decided to involve the user, how the risk was displayed and communicated to the user. The empirical study conducted in [78], which implemented an intensive study on top, used permissions with a high-risk level. They calculated the risk level based on the type of permissions and the probability they will be requested by the app. The risk value for each permission in addition to its technical name and description was transmitted to the users. Although a coloring code was used to indicate the level of risk, still users are involved in technical details which will not help them to take proper decisions regarding the apps' installation. The work presented in [79] has utilized fuzzy logic to measure the risk score. Also, in addition to the permissions and their categories, they took input from different antivirus tools to calculate the score. Their system allowed the user to upload the app's APK through the browser and provided them with a risk report. This report showed the risk score, permission usage rate, and unnecessary permission usage rate in addition to the list of permissions, their categories, and risk level. On the other hand, the authors in [80] have considered the statistical distribution of the Android permissions in addition to the probabilistic functions. The declared but not exploited permissions and vice versa were all considered in their risk analysis. Machine learning was also utilized to measure risk.

In terms of visualizing the permissions and their risks, the authors in [81] introduced Papilio to visualize Android application permissions graphically. This helped them to find the relations among the applications and applications' permissions as well. Papilio was able to find the permissions requested frequently by applications and permissions that either never requested or requested infrequently. The authors in [82] stressed the importance of visualizing the statistical information related to Android permissions. Having graphical representation for the permissions' statics within a specific category encouraged the users to choose more often apps with a lower number of permissions. A privacy meter was used in [83] to visualize the permissions' statistics through a slider bar which outperformed the existing warning system like Google's permission screens. Visualizing app activities enhances user's awareness and sensitivity to the privacy intrusiveness of mobile applications [84]. Another attempt to visualize the permissions statistics was also introduced in [85] using lifelog analysis views in terms of risk history and app's risk view.

From the above-related work, we can observe that the majority of the previous solutions have mainly relied on permissions either statistically or based on probability to analyze Android apps, to classify them as malwares, and to measure their risk level. Although permissions are important to analyze and classify Android applications. However, these permissions should be up-to-date. Also, other important static and dynamic metrics need to be considered to guarantee a comprehensive evaluation and consequently an accurate detection of malware apps.

There have been many types of research on designing malicious detection approaches. Such approaches resort to static analysis of the malware, and others use dynamic analysis, while some methods utilize both static and dynamic analyses to get better detection of a malicious incident. Moreover, the generated datasets will be analyzed in order to detect any potential security threats, regardless whether these datasets were constructed based on static or dynamic tests or even both. Usually, data mining techniques could be used for the purpose of detecting and classifying attacks [42, 52]. Moreover, intelligence techniques could be utilized to even rank the risk by assigning the attack a risk score [42, 86].

The scanning service might fruit in developing a mobile application that is installed on user's devices to examine the Android application and discriminate, if it is a clean app or a malicious app to warn the user and protect her/his Android device. DREBIN [87] is one of the malware detection systems available for smartphones. One of the major features that DREBIN provides is instantaneous malware detection. When a new application is downloaded, DREBIN starts the analyzing process directly. As a result, the user is protected against any unreliable sources. Another example of anti-malware software is HinDroid [88] which has been integrated as one of Comodo's mobile security scanning tools. HinDroid structures the APIs based on heterogeneous information network in order to make predictions about the tested application. Consequently, HinDroid can reduce the time and cost of analyzing Android apps.

5. Statistical analysis

This section presents a statistical study to show the frequency of the used approaches, methods, datasets, and tools in the current systems. Various, related, recent, published solutions in 2017–2018 were considered in this study.

In regard to reverse engineering tools utilized by researchers, APKtool was heavily used by 54% in comparison with other tools (see **Figure 4**). Soot was next with 20% of usage.

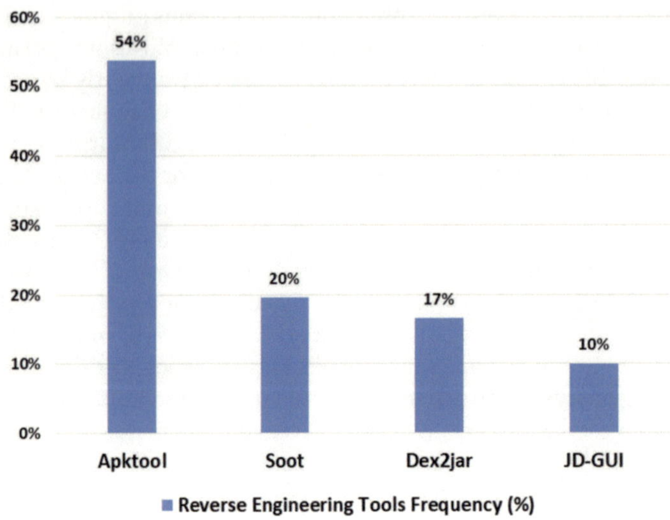

Figure 4.
Reverse engineering tool usage in 2017–2018 research.

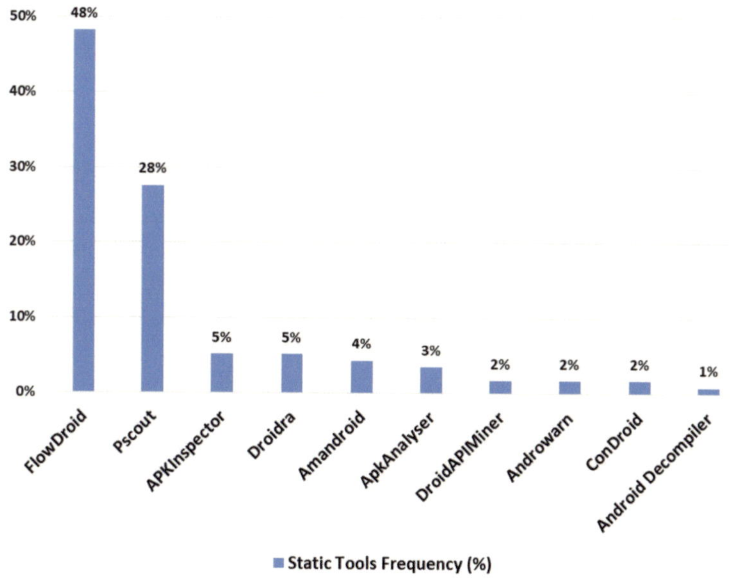

Figure 5.
Static tool frequency in 2017–2018 research work.

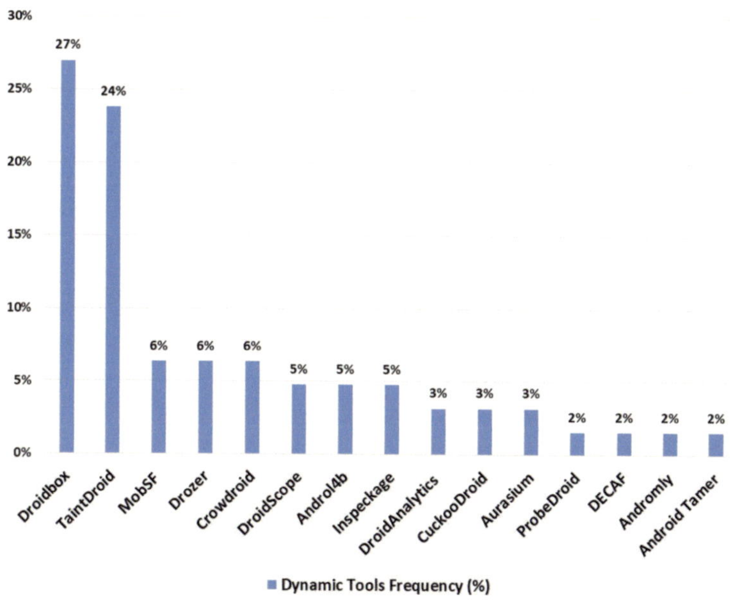

Figure 6.
Dynamic tool frequency in 2017–2018 research work.

Figure 5 shows a comparison among the static tools which were utilized by researchers. It can be observed that 48% of the static-based systems used FlowDroid tool in their solutions. PScout was the second most used with percentage reaching around 28%.

Dynamic analysis tool usage is illustrated in **Figure 6**. The majority of existing solutions used Droidbox with 27% and TaintDroid with 24% in comparison with other approaches. The rest of the results are shown in **Figure 6**.

The results in **Figure 7** reveal that AndroZoo was the most used dataset in 2017–2018. The percentage of usage reached 43%. Genome and DREBIN datasets came next with frequencies 30 and 16%, respectively.

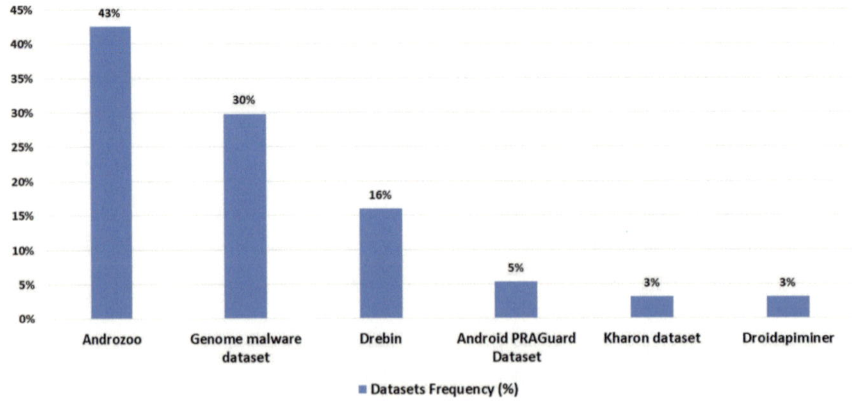

Figure 7.
Dataset frequency in 2017–2018 research work.

6. Conclusions

This chapter highlighted the booming of Android technologies and their applications which make them more attractive to security attackers. Recent statistics of Android malwares and their impact were presented. Additionally, this chapter has provided the main phases required to apply security scanning to Android applications. The purpose is to protect Android users and their devices from the threats of different security attacks. These phases include the way of downloading Android apps, decoding them to generate the source code, and how this code is screened to extract the required features to apply either static analysis or dynamic analysis or both. The feature extraction process resulted in constructing different datasets. Proper data analysis and data mining techniques could be applied to examine the app and classify it as benign or malware with high accuracy. The malware detection service could be implemented and provided in terms of a mobile application that will communicate the scanning results to the user in a friendly way. The chapter was concluded by presenting a statistical study that showed the most used tools and datasets throughout the scanning process for the last 2 years 2017 and 2018.

Acknowledgements

We would like to acknowledge the Security Engineering Lab (sel.psu.edu.sa) team and Prince Sultan University for supporting this work. Special thanks go to Ms. Samah Alsoghyer and Ms. Aala Khayer.

Conflict of interest

The authors declare that there is no "conflict of interest" in regard to publishing this book chapter.

Author details

Iman Almomani[1,2]* and Mamdouh Alenezi[1]

1 Prince Sultan University, Riyadh, KSA

2 The University of Jordan, Amman, Jordan

*Address all correspondence to: imomani@psu.edu.sa

References

[1] Alliance OH. Android overview. Open Handset Alliance. 2011;**8**:88-91

[2] Faruki P, Bharmal A, Laxmi V, Ganmoor GM, Conti M. Android security: A survey of issues, malware penetration, and defenses. IEEE Communications Surveys & Tutorials. 2015;**17**(2):998-1022

[3] Global smartphone sales to end users from 1st quarter 2009 to 2nd quarter 2018 [Internet]. Available from: https://www.statista.com/statistics/266219/global-smartphone-sales-since-1st-quarter-2009-by-operating-system/ [Accessed: 2019-03-22]

[4] Nimodia C, Deshmukh HR. Android operating system. Software Engineering. 2012;**3**(1):10

[5] Brahler S. Analysis of the android architecture. Karlsruhe Institute for Technology. 2010;7(8):3-64

[6] Drake JJ, Lanier Z, Mulliner C, Fora PO, Ridley SA, Wicherski G. Android Hacker's Handbook. New Jersey, USA: John Wiley & Sons; 2014

[7] Mithilesh Joshi BlogSpot. What is android application components and how we use it? 2015. https://mithileshjoshi.blogspot.com/2015/06/ CITATIONS 74 what-is-android-application-components.html [Accessed November 27, 2017]

[8] AppBrain. Number of Android applications on the Google Play store | AppBrain. 2019. Available from: https://www.appbrain.com/stats/number-of-android-apps [Accessed: 2019-03-02]

[9] AV-TEST The IT-Security Institute. Malware Statistics and Trends Report | AV-TEST. The AV-TEST Institute; 2018. Available from: https://www.av-test.org/en/statistics/malware/ [Accessed: 2019-03-02]

[10] GData. Cyber attacks on Android devices on the rise. 2018. Available from: https://www.gdatasoftware.com/blog/2018/11/31255-cyber-attacks-on-android-devices-on-the-rise [Accessed: 2019-04-15]

[11] F-Secure State of cyber security. Available from: https://www.f-secure.com/documents/996508/1030743/cyber-security-report-2017 [Accessed: 2019-04-15]

[12] Security Report 2016/17 [Internet]. Available from: https://www.avtest.org/fileadmin/pdf/security_report/AV-TEST_Security_Report_2016-2017.pdf [Accessed: 2019-04-15]

[13] Delac G, Silic M, Krolo J. Emerging security threats for mobile platforms, MIPRO. In: 2011 Proc. 34th Int. Conv. 2011. pp. 1468-1473

[14] Savage K, Coogan P, Lau H. Security Response – The Evolution of Ransomware. Mountain View (CA): Symantec Corporation; 2015

[15] Liska A, Gallo T. Ransomware: Defending Against Digital Extortion. California, USA: O'Reilly Media, Inc; 2016

[16] (KasperSky) Chebyshev V. Mobile Malware Evolution 2018 [Internet]. Available from: https://securelist.com/mobile-malware-evolution-2018/89689/ [Accessed: 2019-04-16]

[17] Grégio A, Abed R, Afonso V, Filho D, Geus P, Jino M. Toward a taxonomy of malware behaviors. The Computer Journal. 2015;**58**(10):2758-2777

[18] Alenezi M, Almomani I. Abusing android permissions: A security perspective. IEEE Jordan Conference on Applied Electrical Engineering and Computing Technologies (AEECT); Amman, Jordan. 2017

[19] Developer.Android. "Permission Types" [Internet]. Available from: https://developer.android.com/guide/topics/manifest/permission-element [Accessed: 2019-04-15]

[20] Gianluca D, Martinelli F, Matteucci I, Petrocchi M, Saracino A, Sgandurra D. Risk analysis of android applications: A user-centric solution. Future Generation Computer Systems. 2018;**80**:505-518

[21] Seray T, Demetriou S, Ganju K, Gunter C. Resolving the predicament of android custom permissions. In: ISOC Network and Distributed Systems Security Symposium (NDSS). 2018

[22] Arstechnica.com. Your iPhone Calendar isn't Private. 2012. Available from: http://arstechnica.com/apple/2012/06/your-iphone-c

[23] Haining C, Li N, Enck W, Aafer Y, Zhang X. Analysis of SEAndroid policies: Combining MAC and DAC in android. In: Proceedings of the 33rd Annual Computer Security Applications Conference. Orlando, FL, USA: ACM; 2017. pp. 553-565

[24] Ratazzi EP. Understanding and Improving Security of the Android Operating System. No. AFRL-RI-RS-TP-2018-001. New York, USA: Air Force Research Laboratory/Information Directorate Rome United States; 2016

[25] Backes M, Bugiel S, Hammer C, Schranz O, von Styp-Rekowsky P. Boxify: Full-fledged app sandboxing for stock android. In: Proceedings of the 24th {USENIX} Security Symposium ({USENIX} Security 15). 2015. pp. 691-706

[26] Bennet Y, Sehr D, Dardyk G, Chen J, Muth R, Ormandy T, et al. Native client: A sandbox for portable, untrusted x86 native code. In: Proceedings of the 2009 30th IEEE Symposium on Security and Privacy. Berkeley, CA, USA: IEEE; 2009. pp. 79-93

[27] Elenkov N. Android Security Internals: An in-Depth Guide to Android's Security Architecture. San Francisco, California, USA: No Starch Press; 2014

[28] Georgios P, Homburg P, Anagnostakis K, Bos H. Paranoid android: Versatile protection for smartphones. In: Proceedings of the 26th Annual Computer Security Applications Conference. Austin, Texas, USA: ACM; 2010. pp. 347-356

[29] Almomani I, Alkhayer A. Android applications scanning: The guide. In: Proceedings of the IEEE International Conference on Computer and Information Sciences (ICCIS); 3-4 April 2019; Saudi Arabia. Jouf: IEEE; 2019

[30] Allix K, Bissyandé F, Klein J, Traon, L. AndroZoo. In: Proceedings of the 13th International Workshop on Mining Software Repositories—MSR 16. 2016. DOI: 10.1145/2901739.2903508

[31] Abubaker H. Analytics on malicious android applications. International Journal of Advanced Software Computer. 2018;10:106-118. ISSN 2074-8523

[32] Ikram M, Kaafar M. A first look at mobile ad-blocking apps. In: Proceedings of the 2017 IEEE 16th International Symposium on Network Computing and Applications (NCA). 2017. DOI: 10.1109/nca.2017.8171376

[33] Arnatovich L, Wang L, Ngo N, Soh C. A comparison of android reverse engineering tools via program behaviors validation based on intermediate languages transformation. IEEE Access. 2018:12382-12394. DOI: 10.1109/access.2018.2808340

[34] Baskaran B, Ralescu AA. Study of android malware detection techniques and machine learning. In: Proceedings of the Mod. Artif. Intell. Cogn. Sci. Conf. 2016. pp. 15-23

[35] Arshad S, Ali M, Khan A, Ahmed M. Android malware detection & protection: A survey. International Journal of Advanced Computer Science and Applications. 2016;7(2):463-475. DOI: 10.14569/ijacsa.2016.070262

[36] Zachariah R, Yousef M, Chacko A. Android malware detection and prevention. International Journal of Recent Trends in Engineering and Research. 2017;**3**(2):213-217. DOI: 10.23883/ijrter.2017.3028.0uhbl

[37] Baskaran B, Ralescu A. A study of android malware detection techniques and machine learning. In: Proceedings of the IEEE International Conference on Smart Internet of Things (SmartIoT). 2018. DOI: 10.1109/smartiot.2018.00034

[38] Chen J, Wang C, Zhao Z, Chen K, Du R, Ahn G. Uncovering the face of android ransomware: Characterization and real-time detection. IEEE Transactions on Information Forensics and Security. 2018;**13**(5):1286-1300

[39] Arzt S, Rasthofer S, Fritz C, Bodden E, Bartel A, Klein J, et al. Flowdroid: Precise context, flow, field, object-sensitive and lifecycle-aware taint analysis for android apps. Acm Sigplan Notices. 2014;**49**(6):259-269

[40] Au K, Zhou Y, Huang Z, Lie D. Pscout: Analyzing the android permission specification. In: Proceedings of the 2012 ACM Conference on Computer and Communications Security; October 2012. Raleigh, North Carolina, USA: ACM; 2012. p. 217-228

[41] Mujahid S, Abdalkareem R, Shihab E. Studying permission related issues in android wearable apps. In: Proceedings of the IEEE International Conference on Software Maintenance and Evolution (ICSME); September 2018. Madrid, Spain: IEEE; 2018. pp. 345-356

[42] Tao G, Zheng Z, Guo Z, Lyu M. MalPat: Mining patterns of malicious and benign android apps via permission-related APIs. IEEE Transactions on Reliability. 2018;**67**(1):355-369. DOI: 10.1109/tr.2017.2778147

[43] Alenezi M, Almomani I. Empirical analysis of static code metrics for predicting risk scores in android applications. In: Proceedings of the 5th Symposium on Data Mining Applications (SDMA2018); 21-22 March, 2018; Riyadh, KSA

[44] Enck W. TaintDroid: An information-flow tracking system for realtime privacy monitoring on smartphones. In: Proceedings of the ACM Transactions on Computer Systems (TOCS) 32.2. 2014. p. 5

[45] Zhou Y"H. You, get o_ of my market: Detecting malicious apps in official and alternative android markets. In: Proceedings of the NDSS. 2012. pp. 50-52

[46] Grace M. Riskranker: Scalable and accurate zero-day android malware detection. In: Proceedings of the 10th International Conference on Mobile Systems, Applications, and Services. ACM; 2012. pp. 281-294

[47] Security Affairs. DREBIN Android app detects 94 percent of mobile malware [Internet]. Available from: http://securityaffairs.co/wordpress/29020/malware/drebin-android-av.html [Accessed: 2017-12-01]

[48] Arp D. DREBIN: Effective and explainable detection of android malware in your pocket. Proceedings of the NDSS. 2014

[49] Yang W. Appcontext: Differentiating malicious and benign mobile app behaviors using context. In: Proceedings of the Software Engineering (ICSE); 2015 IEEE/ACM 37th IEEE International Conference. IEEE. 2015. pp. 303-313

[50] Yang W. Malware detection in adversarial settings: Exploiting feature evolutions and confusions in android apps. Proceedings of the Proc. ACSAC. 2017

[51] Akhuseyinoglu N, Akhuseyinoglu A. AntiWare: An automated Android malware detection tool based on machine learning approach and official market metadata. In: Proceedings of the 2016 IEEE 7th Annual Ubiquitous Computing, Electronics Mobile Communication Conference (UEMCON). October 2016. pp. 1-7. DOI: 10.1109 /UEMCON.2016.7777867

[52] Allix K, Bissyandé T, Klein J, Traon Y. AndroZoo: Collecting millions of android apps for the research community. In: Proceedings of the 13th International Workshop on Mining Software Repositories—MSR. 2016. DOI: 10.1145/2901739.2903508

[53] Zachariah R, Akash K, Yousef M, Chacko A. Android malware detection a survey. In: Proceedings of the 2017 IEEE International Conference on Circuits and Systems (ICCS). 2017. pp. 238-244

[54] Kaspersky Lab. "Emulator." [Internet]. Available from: https://www.kaspersky.com/enterprise-security/wiki-section/products/emulator [Accessed: 2019-04-01]

[55] Tam K. CopperDroid: Automatic reconstruction of android malware behaviors. Proceedings of the NDSS. 2015

[56] Bhatia T, Kaushal R. Malware detection in android based on dynamic analysis. In: Proceedings of the 2017 International Conference on Cyber Security And Protection Of Digital Services (Cyber Security). 2017. pp. 1-6. DOI: 10.1109/CyberSecPODS.2017.8074847

[57] Wang ZD-ARM. Taming control flow anti-analysis to support automated dynamic analysis of android malware. In: Proceedings of the 33rd Annual Conference on Computer Security Applications (ACSAC'17). 2017

[58] Huang H, Zheng C, Zeng J, Zhou W, Zhu S, Liu P, et al. A large-scale study of android malware development phenomenon on public malware submission and scanning platform. IEEE Transactions on Big Data. 2018: 15-23. DOI: 10.1109/tbdata.2018.2790439

[59] Kaur R, Li Y, Iqbal J, Gonzalez H, Stakhanova NA. Security assessment of HCE-NFC enabled E-wallet banking android apps. In: Proceedings of the 2018 IEEE 42nd Annual Computer Software and Applications Conference (COMPSAC); July 2018. Tokyo, Japan: IEEE; 2018. pp. 492-497

[60] VirusTotal Malware Intelligence Services [Internet]. n.d. Available from: https://www.virustotal.com/learn/ [Accessed 2018-12-15]

[61] Scaife N. Cryptolock (and drop it): Stopping ransomware attacks on user data. In: Proceedings of the Distributed Computing Systems (ICDCS), 2016 IEEE 36th International Conference on. Nara, Japan: IEEE; 2016. pp. 303-312

[62] Tripwire. Early-Warning Ransomware Detection Tool Could Help Protect Users Despite Drawbacks [Internet]. Available from: https://www.tripwire.com/state-of-security/security-data-protection/cyber-security/early-warning-ransomware-detection-tool-could-help-protect-users-despite-drawbacks/ [Accessed 2017-12-02]

[63] Mercaldo F, Nardone V, Santone A. Ransomware inside out. In: Proceedings of the 2016 11th International Conference on Availability, Reliability and Security. ARES; 2016. pp. 628-637. DOI: 10.1109/ARES.2016. 35

[64] Li J, Sun L, Yan Q, Li Z, Srisa-an W, Ye H. Significant Permission

Identification for Machine-Learning-Based Android Malware Detection, IEEE Transactions on Industrial Informatics; July 2018;**14**(7):3216-3225

[65] Mercaldo F, Nardone V, Santone A. Ransomware inside out. Proceedings of the 11th International Conference on Availability, Reliability and Security (ARES); Augest. 2016:628-637. DOI: 10.1109/ARES.2016.35

[66] Sun L, Wei X, Zhang J, He L, Yu PS, Srisa-an W. Contaminant removal for Android malware detection systems. Boston, MA, USA: 017 IEEE International Conference on Big Data (Big Data); 11-14 Dec 2017. pp. 1053-1062

[67] Maiorca D. R-PackDroid: API package-based characterization and detection of mobile ransomware. In: Proceedings of the Symposium on Applied Computing. Marrakech, Morocco: ACM; 2017. pp. 1718-1723

[68] Mercaldo F. Ransomware steals your phone. Formal methods rescue it. In: Proceedings of the International Conference on Formal Techniques for Distributed Objects, Components, and Systems. Heraklion, Greece: Springer; 2016. pp. 212-221

[69] Mercaldo F. Extinguishing ransomware-a hybrid approach to android ransomware detection. In: Proceedings of the 10th International Symposium on Foundations Practice of Security. 2017

[70] Yang T. Automated detection and analysis for android ransomware. In: Proceedings of the High Performance Computing and Communications (HPCC), 2015 IEEE 7th International Symposium on Cyberspace Safety and Security (CSS), 2015 IEEE 12th International Conference on Embedded Software and Systems (ICESS), 2015 IEEE 17th International Conference on. IEEE; 2015. pp. 1338-1343

[71] Gharib A, Ghorbani A. DNA-droid: A real-time android ransomware detection framework. In: Proceedings of the International Conference on Network and System Security. Helsinki, Finland: Springer; 2017. pp. 184-198

[72] Sarma BP, Li N, Chris Gates C, Potharaju R, Nita-Rotaru C, Molloy I. Android permissions: a perspective combining risks and benefits. In: Proceedings of the 17th ACM symposium on Access Control Models and Technologies. Newark, New Jersey, USA; 20-22 June 2012. pp. 13-22

[73] Enck W, Ongtang M, McDaniel P. On lightweight mobile phone application certification. In: Proceedings of the 16th ACM Conference on Computer and Communications Security. Vol. 2009. pp. 235-245

[74] Gates C, Li N, Peng H, Sarma B, Qi Y, Potharaju R, et al. Generating summary risk scores for mobile applications. IEEE Transactions on Dependable and Secure Computing. 2014;**11**(3):238-251

[75] Mathew J, Joy M. Efficient risk analysis for android applications. In: Proceedings of the IEEE Recent Advances in Intelligent Computational Systems (RAICS); 10-12 December 2015. Trivandrum, India: IEEE; 2015. pp. 382-387

[76] Guo C, Xu G, Liu L, Xu S. Using association statistics to rank risk of android application. In: Proceedings of the IEEE International Conference on Computer and Communications (ICCC); 10-11 October 2015. IEEE; 2015. pp. 1-5

[77] Hao H, Li Z, Yu H. An effective approach to measuring and assessing the Risk of android application. Proceedings of the International Symposium on Theoretical Aspects of Software Engineering (TASE); 12-14 Sept. 2015:31-38

[78] Wang Y, Zheng Y, Sun C, Mukkamala S. Quantitative security risk assessment of android permissions and applications. In: Proceedings of the Lecture Notes in Computer Science, LNCS-7964. Newark, NJ, USA: Springer; 2013. pp. 226-241

[79] Yuksel A, Yuksel E, Sertbasa A, Zaim A. Implementation of a web-based service for mobile application risk assessment. Turkish Journal of Electrical Engineering & Computer Sciences. 2017;**25**(2):976-994

[80] Merlo A, Georgiu G. RiskInDroid: Machine learning-based risk analysis on android. In: Proceedings of the IFIP International Conference on ICT Systems Security and Privacy Protection (SEC). 2017. pp. 538-552

[81] Hosseinkhani M, Fong P, Papilio CS. Visualizing android application permissions. In: Proceedings of the Eurographics Conference on Visualization. 2014. pp. 1-10

[82] Kraus L, Wechsung I, Möller S. Using statistical information to communicate android permission risks to users. Workshop on Socio-Technical Aspects in Security and Trust (STAST). Vienna, Austria: Co-located with 27th ieee computer security foundations symposium (CSF); 18 July 2014. pp. 1-9

[83] Kang J, Kim H, Cheong YG, Huh JH. Visualizing privacy risks of mobile applications through a privacy meter. In: Information Security Practice and Experience. Lecture Notes in Computer Science. Vol. 9065. Beijing, China: Springer; 2015. pp. 548-558

[84] Eze C, Nurse J, Happa J. Using visualizations to enhance users' understanding of app activities on android devices. Journal of Wireless Mobile Networks, Ubiquitous Computing, and Dependable Applications. 2016:39-57

[85] Yoo S, Ryu H, Yeon H, Kwon T, Jang Y. Personal visual analytics for android security risk lifelog. In: Proceedings of the 10th International Symposium on Visual Information Communication and Interaction. 2017. pp. 29-36

[86] Dash S, Suarez-Tangil G, Khan S, Tam K, Ahmadi M, Kinder J, et al. DroidScribe: Classifying android malware based on runtime behavior. In: 2016 IEEE Security and Privacy Workshops (SPW). 2016. DOI: 10.1109/spw.2016.25

[87] Arp D. Drebin: Effective and explainable detection of android malware in your pocket. In: Proceedings of the 2014 Network and Distributed System Security Symposium. 2014. DOI: 10.14722/ndss.2014.23247

[88] Hou S. HinDroid. In: Proceedings of the 23rd ACM SIGKDD International Conference on Knowledge Discovery and Data Mining—KDD 17. Halifax, NS, Canada: ACM; 2017. DOI: 10.1145/3097983.3098026

Section 2

Next-Generation Optical Access Technologies

Chapter 4

Deployment of PON in Europe and Deep Data Analysis of GPON

Tomas Horvath, Petr Munster and Josef Vojtech

Abstract

This chapter discusses the extensibility of fiber to the x (FTTx) households, specifically in the territory of the European Union. The Czech Republic has made a commitment to other member states to provide connectivity of at least 100 Mbit/s for half of the households by 2020. Although Internet access in the Czech Republic is mostly dominated by wireless fidelity (WiFi), this technology is not capable of meeting the demanding current demands at a reasonable price. As a result, passive optical networks are on the rise in access networks and in mobile cell networks by fiber to the antenna (FTTA). Passive optical networks use much more complex networks. In cooperation with Orange Slovakia, the analysis of the transmitted data was conducted. The optical network unit management and control interface (OMCI) channel data, as well as the activation data associated with specific end units, were analyzed. We propose a complete analysis of the end-unit-related activation process, download, and initialization of the data image for setting the end units and voice over Internet protocol (VoIP) parameters. Finally, we performed an analysis of the transmission of dying gasp messages.

Keywords: dying gasp, GPONxpert, OMCI channel analysis, ONU activation process analysis, PON deployment, transmission convergence layer

1. Introduction

The optical infrastructure is essential for current applications that demand a high bandwidth [1–3]. The International Telecommunication Union (ITU) has been developing standards for passive optical networks (PONs) for over 20 years [4, 5]. The second most active organization in this area is the Institute of Electrical and Electronics Engineers (IEEE) [6, 7].

Passive optical networks are currently expanding, as the European Union (EU) has allocated budget to extend the coverage of these networks [8]. Today, the access network is not only about transferring data streams from/to the Internet. The popularity of Amazon TV, Netflix, and so on puts increased demands on bandwidth. Current transmission speeds are not sufficient, and a bandwidth of at least 100 Mbit/s in every household is still under consideration. In the Czech Republic, the utilization of gigabit PON (GPON) standard still dominates. However, such standard was in its first version approved back in 2003 [9]. This standard makes it possible to achieve a bandwidth of up to 2.5 Gbit/s in full duplex mode, but the disadvantage is that the bandwidth is fully shared by all end users (in theory, up to 128 customers per port). The available bandwidth can be operatively changed in time and according to the requirements using dynamic bandwidth allocation (DBA) algorithms [10–13]. The decreasing cost of the necessary devices allows GPON optical line termination (OLT) to be used more often for service providers; on the other hand, the standard in use may not be sufficient for the future. The cost of the next-generation PON (XG-PON) terminal units is still quite high, regardless of the OLT unit price. The price of the technology itself is determined by the price of the optical network unit (ONU) terminal units. The advantage of deploying next-generation networks would be the ability to a share the transfer rate of up to 10 Gbit/s. Together with appropriate DBA algorithms, the full bandwidth utilization or its adequate distribution between endpoints would be efficiently used. GPON networks theoretically allow us to transfer data up to 19 Mbit/s for each ONU (considered for the maximum transfer rate and a split ratio of 1:128). XG-PON networks are limited by higher split ratios but have higher transfer rates available. Theoretically, 39 Mbit/s can be achieved for each ONU. In other words, the transfer rates are the maximum possible in both GPON and XG-PON networks. Usually, the guaranteed transfer rates are several times lower according to the use of a transmission container (T-CONT) [14].

2. Current state of the access networks in the Czech Republic

The Czech Republic has committed itself within the European Union to ensuring a transmission rate of at least 30 Mbit/s toward users by 2018 [15]. In 2020, the next milestone is going to be to increase the downlink speed up to 100 Mbit/s for approximately half of all households [15]. Both variables account for asymmetric transmission rates (usually higher transmission rates in the downstream direction). Based on [16], this "scarcity" should be eliminated by 2030. At that time, only a symmetric variant of Internet access will be considered.

Current technologies such as asymmetric digital subscriber line (ADSL) are no longer able to meet the previously mentioned bandwidth requirements. The plans of the Czech Republic include a GPON or a variant of an active optical network. As presented in [17], the formal definition for next-generation networks can be defined as follows: next-generation networks (NGNs) are networks based on data packet transfer technologies capable of providing electronic communications services, allowing for the use of various high technologies that are able to manage and control the quality of the provided services, and whose functions related to these services are independent of basic transmission technologies. The network provides subscribers with unlimited access to various providers of publicly available electronic communications services and consistently supports the provision of services to subscribers at any point in the network. Additionally, next-generation networks can be split into backbone and access networks. This work, however, deals exclusively with access networks.

On the other hand, the Czech Republic is not entirely prepared to satisfy the high demands on the connection speed in all locations. Based on [18], the dominant transmission rates were mostly up to 10 Mbit/s. No significant growth of higher transmission rates has been recorded.

In 2016, the Czech Telecommunication Office published an annual report summarizing current technologies for Internet access. The associated graph can be seen in **Figure 1**. **Figure 1** clearly shows that the dominant technology in this area is wireless fidelity (WiFi) (26.8%), i.e., wireless transmission of information. The annual report does not include the frequencies used; however, the basic frequencies in the

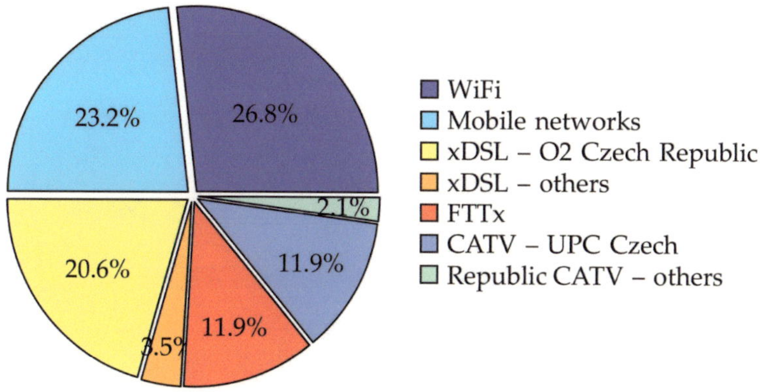

Figure 1.
Access technologies market share in the Czech Republic [15].

license-free band (2.5/5 GHz) can be assumed. The second technology with the highest penetration is represented by mobile networks (23.2%). The third technology combines all types of xDSL technologies. This area is dominated by Telefonica O2 Czech Republic, a.s., with a penetration of 20.6%. Other xDSL technologies only reach 3.5%. According to [16], fiber to the x (FTTx) connections at the same value of penetration (11.9%) as cable operator UPC Czech Republic, s.r.o., currently offers the fastest connection speed of 500/30 Mbit/s (depending on the location). Conversely, FTTx connections depend only on the selected standard as with the fiber to the home (FTTH) variant. FTTx connections can support transmission rates up to 10/10 Gbit/s (depending on the number of end units connected to the OLT control unit).

The properties of the next-generation access networks can be summarized as follows [16]:

- providing high transmission rates for subscribers and providing reliable services through optical networks or other comparable technologies,

- supporting a variety of advanced digital and converged services based on Internet protocol (IP),

- providing significantly higher transmission rates in the downstream direction, i.e., toward the user.

3. Household penetration

The current state of fiber to the building (FTTB) or FTTH connections is generally problematic to analyze. These data are usually not freely available, and the cost of these documents is high (on the order of thousands of dollars). A company named IDATE has published its market research for the FTTH Council Europe conference [19]. The outcome of the analysis for Europe clearly shows that Latvia has the best FTTB/H (households) connection (see **Figure 2**). Their household penetration is approximately 50.6% (25.3% are FTTH connections). Another dominant country is Sweden, with a total penetration of 43.3% (only 8.5% are FTTH connections). The total penetration for the Czech Republic is very low compared to other countries, with a total penetration of 3.7% (only 2% are FTTH connections). Compared to the neighboring state, Slovakia has an overall penetration of 17.7%

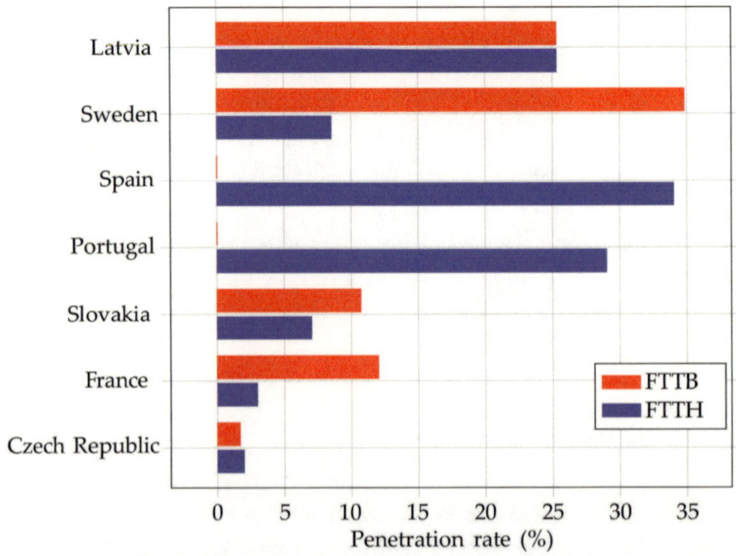

Figure 2.
Selected EU states with FTTH/B penetration [19].

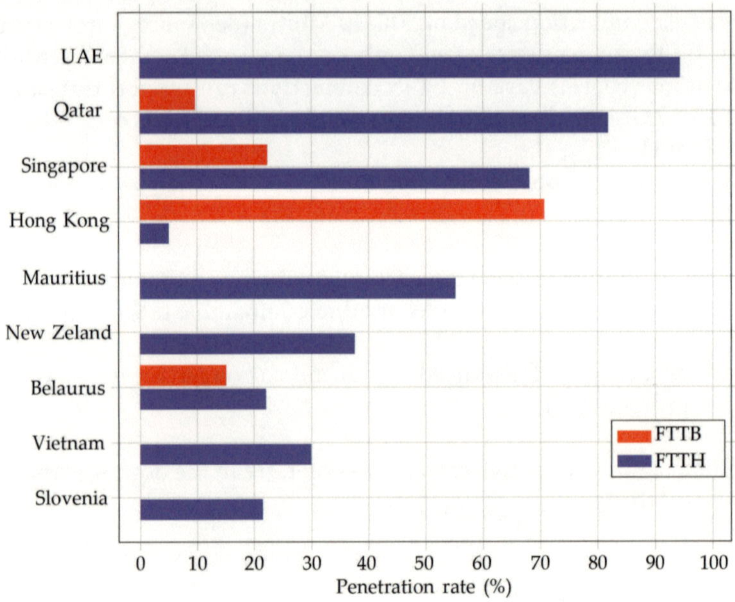

Figure 3.
Selected global states with FTTH/B penetration [19].

(7% are FTTH connections, and the remaining 10.7% are FTTB connections). This is mainly because in Slovakia, there is a very strong operator, Orange SK. Orange SK may test the use of new technologies in this relatively small market, and if this technology stands, it can be deployed, for example, in Orange home (formally France Telecom) in France.

Another objective of the current FTTB/H connection analysis is to focus on the global market (see **Figure 3**). Globally, the United Arab Emirates (UAE) has a total penetration of 94.3%. This penetration is completely composed of FTTH

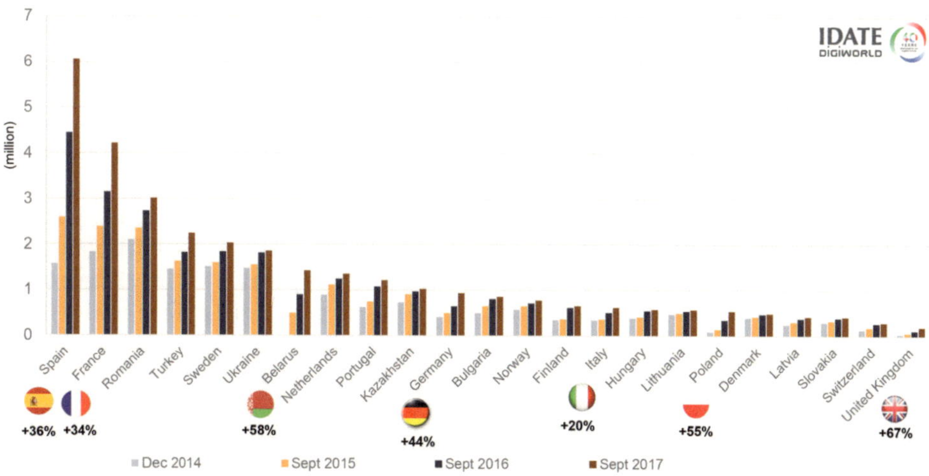

Figure 4.
Progress in FTTH/B deployments according to IDATE [19].

connections. Strong competitors for the UAE are Qatar and Singapore. Qatar has a penetration of 81% for FTTH connections and of 9.4% for FTTB connections. Singapore has a similar total penetration, but in a different ratio, 68% for FTTH connections and 22.3% for FTTB connections. FTTH-only countries are Mauritius, New Zealand, Spain, Vietnam, Portugal, Slovenia, Jamaica, Saudi Arabia, Australia, Macedonia, Switzerland, Oman, Kuwait, Chile, Ecuador, Colombia and Angola. The total penetration of the last 4 states does not exceed 5% [19].

The total penetration is strongly dependent on the number of individual connections. IDATE focused on the analysis of the global market and the comparison of the state of connections in buildings in four stages, December 2014, September 2015, September 2016, and September 2017 (see **Figure 4**). The largest increase in connections was in Poland, with a total difference of 46%. Italy was the second country with the largest increase in connections (35%), followed by Great Britain and France (31%), Spain (24%), and Portugal (22%). Unfortunately, the Czech Republic was not included in this analysis because the number of connections is not as significant. In other words, the trend of building connections is greater in Belarus, Norway, Lithuania, and Hungary.

A detailed view of the number of FTTB/H customers can be seen in report [20]. The report shows that at the end of 2010, the total number of customers was balanced across the EU28 and the commonwealth of independent states (CIS). From a wider perspective, the EU39 reached approximately 8 million customers. However, this difference must be attributed, in particular, to 11 other countries that are counted in the EU39. The aligned trend between the EU28 and the CIS was maintained until 2015. Later, the number of customers increased in the EU28, and the previous dominance of the CIS was diminished. In September 2017, the total number of customers was approximately 25 million, while for the CIS "only," it was 20.5 million. Most places for customers are connected to the provider's network, but there have also been new locations for housing, new towns, and satellite residences created. During the preparation of the work, developers are working hard to build a data infrastructure and negotiations are taking place between Internet services providers (ISPs) and developers. EXFO defines these connections as home passed: premises to which an operator has the capability to connect in a service area, but the premises may or may not be connected to the network [21].

4. Access networks and 5G networks

The primary determination of all technologies for xPON is evident from their name, a passive optical (access) network. This trend continues from the original asynchronous transfer mode PON (APON), broadband PON (BPON), GPON, XG-PON, and the latest approved next-generation PON stage 2 (NG-PON2) recommendations. The latest recommendation has become the pioneer of extending the passive optical network to mobile customers as well. However, residential customers with a fixed connection (flat or house) still remain the priority. With the onset of 5G technology in mobile communications, it will be necessary to reduce the area of cells to ensure coverage of the entire territory by radio signals. This is mainly due to the increasing permeability and diminishing cell size, so it is necessary to build more cells that cover the same area. It is possible to divide the area according to its antenna density into low density (<20 small cells/km^2), medium density (<75 small cells/km^2), dense (<200 small cell/km^2), and ultrahigh density (>200 small cells/km^2). Current long-term evolution (LTE) technology has been providing broadband data services; however, these technologies seem to be inadequate for certain services (virtual reality or generally the most sensitive services for low latency, such as access to data networks of the Internet of things devices). Current customer needs may include gigabit transmissions per second, smart home/buildings, self-driving car, working and playing in the cloud, and 3D or UHD video. Minimal latency requirements will be determined mainly based on data transmission within the national network (10–200 km). The transmission delay in the current networks ranges from 5 to 41 ms, and the delay for the access part of the network (1–10 km) is approx. 7–12 ms. Another key factor that affects the delay is the time it takes to process incoming requests from a data center (approximately 8 ms). The round-trip time (RTT) of current networks is approximately 106 + 8 ms. 5G networks aim to limit this value to 14 + 8 ms. The major merit of RTT depreciation will be to move cloud services closer to the user. Then, the RTT will be reduced to 14 ms, which will primarily generate a delay (7 ms) on the access technology. However, the question remains how the operators will move the data centers closer to the customer, since until now, a distance of 200 km a data center from the customer has been enough. Such a distance is not sufficient for 5G networks.

Among the available technologies covering the 5G signal area, there are technologies for access networks: G.fast, data over cable service interface specification (DOCSIS) and NG-PON2. G.fast technology offers symmetric transmission speeds of up to 500 Mb/s over a short distance (up to 100 m). This speed can be increased to 10 Gb/s, but the overall system reach will be shortened. In theory, G.fast can only be deployed in special cases, such as brownfield scenarios, to ensure connectivity of very small cells in buildings. The basic prerequisite is the combination of functions within the baseband unit (BBU) and remote radio unit (RRU). DOCSIS 3.1 offers bandwidth of 10/1–2 Gb/s share per coaxial segment (192 MHz orthogonal frequency-division multiplexing (OFDM) channels). Full-duplex communication (current downstream and upstream) can take up to 10 Gb/s per coaxial segment. However, neither of these methods is capable of fully serving the 5G network because the available bandwidth is shared and the common public radio interface (CPRI) does not support the lowest possible latency for transmission.

The basic idea behind the NG-PON2 network is to provide all end stations with sufficient bandwidth. The station shares the total bandwidth that the associated OLT unit is able to handle properly. NG-PON2 network parameters such as distribution ratios, power levels, transfer rates, etc. are described in [22–25]. In 5G network areas, there is ultradense deployment of basic radio stations required,

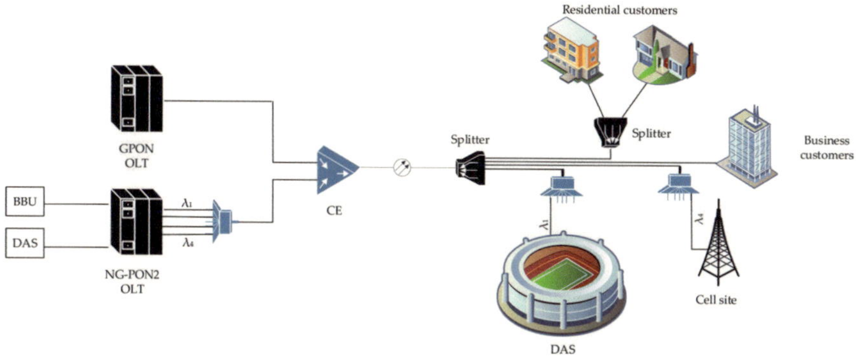

Figure 5.
Coexistence NG-PON2 and GPON scheme with dedicated lambdas for 5G networks.

and their radiations are constrained to prevent intra- and inter-cell interference. In general, the reach of NG-PON2 (up to 20 km from the OLT) is sufficient for covering an acceptable number of end users and for effective usage of its coverage (the division of covered territory into several smaller sectors/cells). The use of access technologies for data transfers or generally for triple play has already been noted out by ITU in [26]. **Figure 5** defines a possible scheme of the NG-PON2 network for its connection to the 5G network. The connection can be realized by dedicated wavelengths (λ). By using a coexistence element (CE), such a coexistence scheme for older PON standards under the ITU recommendations can be established. Regarding the aforementioned dedicated wavelengths, up to 4λ with a 10 Gbit/s transfer rate is considered. One disadvantage of this radio tower connection method is the custom lock method that is publicly available but is much more complex than in the case of the IEEE network. As a result, it will be necessary to use the conversion station to transmit the signal from the radio station toward the end customers.

5. GPON frame structure and activation process analysis

At present, GPON is one of the most promising solutions for modern access networks. Among other useful and important features, it provides us with triple play services on a single optical fiber, good scalability, DBA, simple topology management, etc. In comparison with the previous standards that only supported transmission over asynchronous transfer mode (ATM), GPON is the first standard that supports transmission over both ATM and ethernet technologies. In the ethernet mode, the ethernet frames are encapsulated using GPON encapsulation mode (GEM) and transferred inside GEM frames. As a result, some ethernet structures, such as interpacket gap, preamble, or start of frame delimiter, are not available. For more information, see **Figure 6**.

The basic GPON topology comprises the following three components: OLT, ONU, and optical distribution network (ODN). Typically, there is/are a single/more OLT/s in the network (depending on the preferences of the associated Internet service provider) performing encapsulation and de-encapsulation of downstream and upstream network traffic, respectively, for multiple end users (up to 128 end users per port). The ONU is located at the end user's premises and converts the signals from the optical to the electrical domain. Finally, an ODN is composed of the elements placed between OLTs and ONUs such as optical fibers, splitters, and connectors.

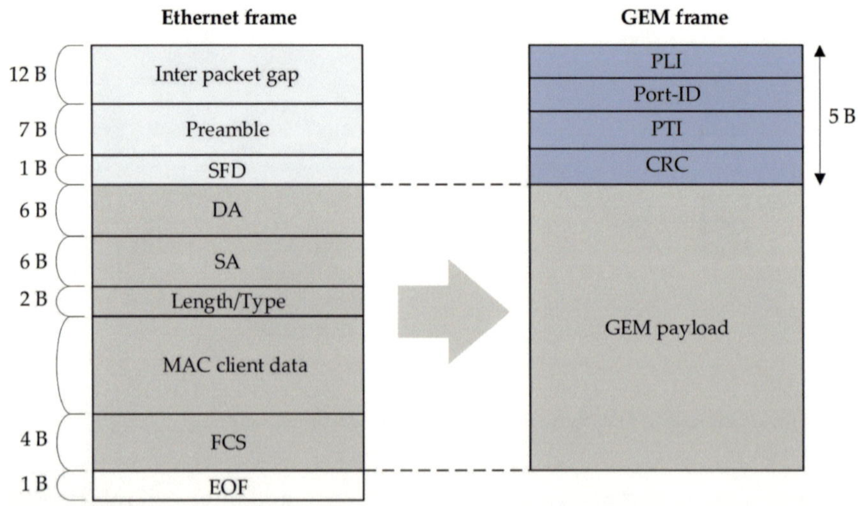

Figure 6.
Ethernet encapsulation into the GEM frame [26].

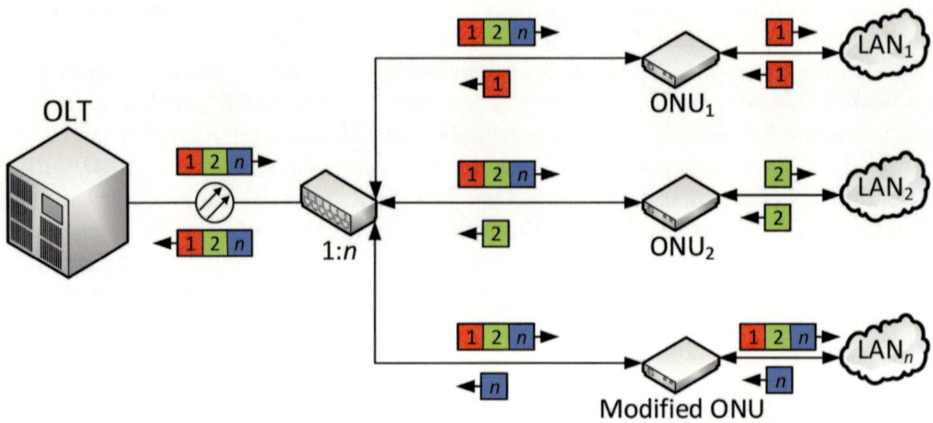

Figure 7.
Interception of downstream communications.

The risk of passive interception of communications results directly from the nature of PON communication. Downstream communication can be secured; however, the major disadvantage is that security is only optional. A potential attacker could, therefore, modify the firmware of an ONU and eavesdrop on all the communication in the downstream direction [26, 27]. The traffic in this direction can also be captured using optical radiation detectors, not necessarily an ONU detector, so encryption of data in the downstream direction had to be introduced [28]. However, the subsequent processing of the captured signal is an essential next step. The situation where the modified end unit receives all frames, including those not directly assigned to it, can be seen in **Figure** 7.

The previously mentioned passive interception could also occur in the upstream direction because no security is used for the upstream communication. This type of interception is complicated; however, it is feasible. The recommendations for use do not define any security for this direction of communication. The reason for this is based on the fact that it is not possible to capture the communication of other end users in the upstream direction via the ONU, so communication is not necessary to be encrypted. To eavesdrop on the communications in this direction, a potential attacker would have to disrupt the PON optical line. This situation would, however, affect the transmission properties of the network in question, which should be captured by the service provider's surveillance center. This way of interception is therefore very unlikely [29].

The abovementioned reason resulted in the fact that no security standard has been provided for any of the individual PON standards. In the event of encryption of the downstream transmission, e.g., using advanced encryption standard (AES) or other secret key-based technology, these keys would have to be sent in an unsecured form—plain text in the upstream direction. It was based on the assumption that upstream communication was safe; therefore, it was not necessary to provide any additional security [30].

The research described in [31] focused specifically on the possibilities of interception of the communication in the upstream direction. The authors tested whether it was possible to intercept the communication through the back reflections of the optical signal. These reflections could be caused by a variety of commonly used optical components, such as passive optical hubs and/or connectors. Moreover, the optical positive-intrinsic-negative (PIN) detectors and avalanche photodiode (APD), as well as the preamplifiers, also had an effect on capturing the communications in the upstream direction. Testing was carried out at various ODN configurations, mainly aimed at testing the back reflection of the optical signal. The success of the potential attacker depended primarily on the type of connector used and the photodetector. A polished connector (PC) was considered inappropriate in terms of network security. The angled polish connector (APC) reduced signal reflections by virtual vertical grinding. Using an APD connector, however, increased the probability of a successful interception of the communicating ONU. Nevertheless, the capability of eavesdropping in the upstream direction was not dependent on the particular bit rate; it depended mostly on the power level of the retroreflection and the type of connector in use [31].

The following demonstrates how to intercept communication in both directions with a specialized tool in hand. Real-time network analysis of the transmitted data (ONU management and control interface (OMCI) channel and GEM data units for end units) was performed. For the purpose of the demonstration, the GPONxpert tool was used. This tool has been developed specifically for passive optical networks. The tool allows for the real-time analysis of ONU-ID, performance levels, and Alloc-ID. However, a detailed analysis of the transmitted data is still necessary to be implemented in the form of postprocessing. Although the manufacturer, TraceSpan, also has other modifications to this device, for our purposes, the most popular measuring device was used. The lite versions contained support for ONU-ID analysis. The real-time analysis of levels, Alloc-IDs, and other parameters was stored using field programmable gate array (FPGA) and sent to the device manufacturer for the postprocessing. The manufacturer then sent the report from the measurement back to the customer.

This work is focused on the analysis of downstream and upstream transmission in GPON standard topology. At the start of the measurement, all ONUs search for their associated network parameters (e.g., serial number, ONU-ID, etc.) that are stored inside the previously mentioned GEM frames. Since the distance between the ONUs and the OLT are different, it was also necessary to use an equalization delay parameter that is assigned by the OLT during the activation process. For more information, see [31, 32]. Consequently, all ONUs wait for a random period prior to starting data transmission. In the frame of this work, data are broadcasted in the downstream direction. In the upstream direction, time slots assigned by the OLT are used instead. Moreover, in this work, we did not use the DBA algorithm. Consequently, all ONUs are expected to transfer data in time slots with prespecified start and stop times.

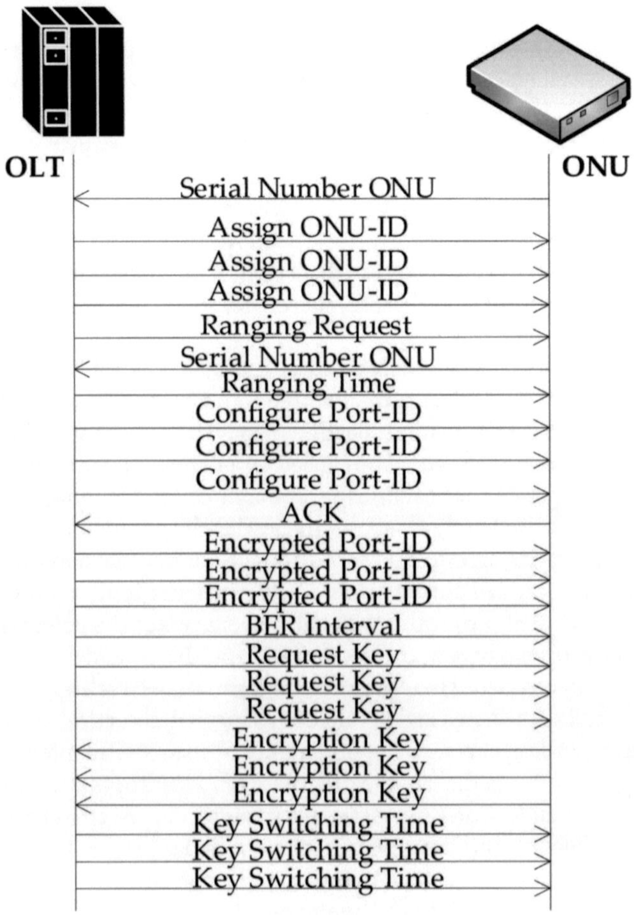

Figure 8.
GPON activation process with encryption channel establishment messages [29].

To summarize, on the one hand, this work is interested in the analysis of user data and the activation process. However, on the other hand, the description of the activation process is omitted, as has already been described in our previous work [32]. Since the user plane and control plane data are transferred using GEM frames, it is not possible to use a common packet analyzer such as Wireshark. For this purpose, we used a GPONxpert analyzer in a standalone mode in which all data are transferred and saved to a hard drive. Therefore, to perform a deeper inspection or analysis, all the data must to be postprocessed. In general, the control plane data can be divided into signaling, OMCI. First, we focused on the signaling data analysis. When the connection is established, messages such as Assign ONU-ID, Configure Port-ID, Assign Alloc-ID, Encrypted Port-ID, Encryption_key, key_request_message, and Key_switching_time are transmitted three times. This, as well as a complete GPON signalization, can be seen in **Figure 8**.

It can be seen that a physical layer operations, administrations and maintenance (PLOAM) message, specifically the "Serial number ONU," are transferred from the ONU to the OLT. This message holds information such as the vendor serial number, a list of supported data profiles, and the value of random delay of 82 μs [28]. The OLT uses these messages to extract the serial number and allocate the associated ONU-ID. Moreover, to minimize the impact of unequal distances among the ONUs and the OLT, it uses unique random delays for each of the ONUs that are based on the time between two successive "Serial number ONU" messages. As soon as the

OLT receives the ONU-ID, it sends the PLOAM message: "Assign ONU-ID." At this point, even though the OLT is aware of the assigned ONU-ID, it is not able to use unicast addressing because the ONU itself still cannot recognize the ONU-ID as its own, and therefore, broadcast addressing needs to be used (the ONU serial number is taken as the identifier) [29]. This means that every ONU receives this message; however, based on the comparison of the incoming and internal serial numbers, only the targeted ONU processes the message. In **Table 1**, it can also be seen that ZTE company is the final unit manufacturer. Based on hard-defined bytes in the MAC address, the manufacturer can be checked directly using its unique label: "0xC03B4EB4." GPON networks supported the transfer of ATM cells; however, in the last review in 2014, this support completely disappeared as these networks did not find their real application. For this reason, "ATM support Disable" can also be observed in the captured data. On the other hand, GEM support is necessary for any GPON data transfer: "GON support Enable." The captured data also have a description of the signal's power level, however, only with the following levels: low/medium/high power.

After the OLT sends the "Assign ONU-ID" message, it consequently sends the "Ranging request" message using the specific ONU-ID. Consequently, the ONU is capable of using a single grant to transmit data. The OLT unit's response to the "Serial Number ONU" message is a PLOAM message, "Assign ONU-ID." This message already carries a unique identifier for the designated end unit. From the nature of PON technology, it is clear that each end unit receives all messages. Using the unique ONU-ID, also called a serial number (if ONU-ID is not assigned), ONUs decide which messages to process. In this case, the assignment of ONU-ID = 1, i.e., the first end unit has already been replied to. The serial number of the unit equals "0x5A544547C03B4EB4", the Psync field is fixed and does not change throughout the communication. This fact is evidenced by the other messages listed in **Table 1**. "Ident Superframe Counter: 499314877" specifies the order of the transmitted frame/s. The ONU endpoint activation process in the GPON network is based on the sending of specific messages three times in a row. The second copy of the message is left for the demonstration of the Superframe counter being incremented by 1. After that, the ONU responds with the "Serial number ONU" message using the maximum priority T-CONT class (i.e., urgent data). The OLT computes a new value for the equalization delay using the "Ranging Time" message sent by the ONU. In the initial ONU report, the unit generates a random delay of 82 μs. The control unit must virtually ensure the same distance for all ONU end units. Each unit is located at a different distance, different customer stores, and/or residential units or streets. Supporting up to 20 km in the distribution part allows for the entire housing estate to be connected. The OLT sends a "Ranging request" message to specify a unique ranging time for each ONU. For this particular message, ONUs are required to respond immediately with their ONU-IDs and serial numbers. The OLT unit repeats the "Ranging request" message three times in total. It is important to note the second response, where the ONU specifies the mandatory parameters such as ONU-ID, the serial number (now omitted), and adds information about the Urgent PLOAM waiting and Traffic waiting in type 2 T-CONTs. The individual T-CONTs represent the distribution of traffic according to their classification by importance. T-CONT 1 responds to urgent data, i.e., data with the highest priority (e.g., voice over Internet protocol—VoIP) and fixed bandwidth. TCONT2 + 3 transfer Internet protocol television (IPTV) data with guaranteed bandwidth, T-CONT 4 is commonly used for best-effort data, and the last T-CONT5 is a mixed type including all types of bandwidth and services. Based on the received OLT responses, the OLT unit evaluates the assigned delay for the given ONU and sends the delay value

ID	ONU-ID	Message type		Message type
2	Unassigned ONU ID	Serial number ONU	Vendor ID: ZTEG, Vendor SN: 0xC03B4EB4, Random Delay: 82 µs, ATM Support: Disable, GEM support: Enable, ONU TX power level: high power	PLOAM
117	Broadcast message	Assign ONU-ID	ONU ID: 1; serial number: 0x5A544547C03B4EB4; Psync: 0xB6AB31E0; Ident Superframe Counter: 499314877; PLOAM CRC: 142	PLOAM
118	Broadcast message	Assign ONU-ID	Ident superframe counter: 499314878	PLOAM
120	1	Ranging request	Psync: 0xB6AB31E0; Ident FEC Indicator: 1; Ident Superframe Counter: 499315777	BWmap
1	1	Serial number ONU	ONU ID: 1; vendor ID: ZTEG; vendor SN: 0xC03B4EB4; random delay: 0	PLOAM
121	1	Ranging request	Psync: 0xB6AB31E0; Ident FEC Indicator: 1; Ident Superframe Counter: 499315777	BWmap
2	1	Serial number ONU	Delimiter: 0xAB5983; ONU ID: 1; Urgent PLOAM waiting: 1; Traffic waiting in type 2, 3, 4, 5 T-CONTs: 0	PLOAM
122	1	Ranging time	Path EqD descriptor: main path EqD; delay: 265409	BWmap
125	1	Request password	Ident Superframe Counter: 499318309	PLOAM
1	1	Password	Password (Hex): 0x47433033423445423400; password (ASCII): GC03B4EB4	PLOAM
126	1	Request key	Psync: 0xB6AB31E0	PLOAM
4	1	Encryption key	Key index: 0; fragment index: 0; key bytes: 0x681A055363E86213	PLOAM
7	1	Encryption key	Key index: 0; fragment index: 1; key bytes: 0x62677982F890BA9C	PLOAM
127	1	Key switching time	Superframe counter: 499321133	PLOAM
10	1	Acknowledge	DM_ID: key switching time	PLOAM
130	1	Configure Port-ID	Activate: enable; port-ID: 1	PLOAM
13	1	Acknowledge	DM_ID: configure port-ID	PLOAM
133	1	Encrypted Port-ID/VPI	Port-ID: 1	PLOAM
16	1	Acknowledge	DM_ID: encrypted port-ID/VPI; ONU ID: 1	PLOAM
136	1	BER interval	BER interval: 40000	PLOAM
19	1	Acknowledge	DM_ID: BER interval	PLOAM
142	1	Assign Alloc-ID	Alloc-ID: 1; Alloc-ID: Type GEM payload	PLOAM
22	1	Acknowledge	DM_ID: assign Alloc-ID	PLOAM

Table 1.
Activation process details in captured data in real GPON networks.

to the "Ranging time" message. GPON networks support so-called backup paths and link recovery systems when an alternative route is available. The message contains two fields: "Path EqD Descriptor: Main Path EqD" identifying the primary path and the backup path (the backup path was not available at the time of testing; therefore, it is not included in the message). The delay value specifies the delay for the end unit in "Delay: 265409," but this value does not match the value in μs. These steps set the basic communication parameters, the assigned ONU-ID, and the equalization delay. During the measurement, secure communication was enabled. The definition of reached states in which communication security can be performed and the prerequisites for negotiating the key are given in [33–35]. The entire process is started with the PLOAM message, "request password" containing "Ident Superframe Counter: 499318309." This message requires the end unit to respond with the same message with a password three times in a row. The captured data contain two fields: "Password (Hex): 0x47433033423445423400" and "Password (ASCII): GC03B4EB4." Next, the "Request Key" message is sent, the content of the message is not fully defined in this case; it is necessary to respond to this message with the Encryption Key message. The "Encryption Key" message consists of "Key Index: 0," "Fragment Index: 0" and "Key Bytes: 0x681A055363E86213." The sequence of these messages is followed and sent three times in a row. In our case, a single message is not enough to deliver the key, so another three messages are used to deliver the remaining part of it. This fact is illustrated by the following: "Fragment Index entry: 1," and "Key Bytes: 0x62677982F890BA9C." The next "Key Switching Time" message should define the start time when a new key is used that was not reached because the tool did not detect these fields. It only detected "Superframe Counter field: 499321133." The start time field contents must confirm the end unit using the "Acknowledge" message. The "Acknowledge" message contains the "Downstream Message Id: Key switching Time" field, confirming the previous message. Next, the OLT sends the "Configure Port-ID" message to the ONU specified by the ONU-ID. In the context of data transmission, the ONU-ID is used for the data flow allocation in a GEM frame. The ONU had to send the acknowledgement (ACK) messages three times (one for each of the received messages). As visualized in **Table 1**, the downstream message identification (DM_ID) contains a "Configure Port-ID" field that holds the confirmed message's name, and an ONU ID equaling the ONU-ID of the end unit (in our case 1). Subsequently, the OLT checks whether the Port-ID is encrypted. If it is not (i.e., the ONU remains in the registration process), the ONU sends the ACK message as a response to each correctly received message. Next, the OLT sends a "BER" (Bit Error Rate) message to specify an accumulation interval for each of the ONUs (number of downstream frames per ONU) that is used to count the number of downstream bit errors [29]. At this point, the ONU knows the Port-ID. However, to establish bidirectional data communication, the Alloc-ID is required to identify a traffic-bearing entity (e.g., T-CONT), which represents the recipient of the upstream data allocated during the BWmap procedure [29]. It is important to note that each ONU requires at least a single Alloc-ID that is equal to the ONU-ID and that is not transmitted by the OLT in the "Assign Alloc-ID" message. In this work, the following Alloc-ID was provided by the OLT: 1. The end unit must always contain at least one ONU-ID identifier, but it may contain several Alloc-IDs. Often, the initial Alloc-ID corresponds to the assigned ONU-ID, which also occurred in this case. The ONU acknowledges each of the PLOAM messages. After that, the encryption of the Port-IDs is rechecked. Nevertheless, it should be mentioned that data encryption is optional, and in reality, many ISPs do not use Port-ID encryption.

5.1 OMCI channel analysis

After the signaling phase is over, the operation, administration and maintenance (OAM) can be transferred using the OMCI channel. In our work, the OMCI procedures begin when the OLT sends a "Get/Set request" message to the ONU. When the ONU receives such a message, it responds with its own "Get/Set" message. In this work, we used a single ONU, see **Table 2**. At this point, the crucial phase of the OMCI analysis is the software image entity type inspection, as the ONU is to be authorized by its own serial number against the database of the OLT (depending on the particular ISP implementation). In the case that the OLT does not have the record of the ONU in the database, the ONU is not allowed to download the software image along with the configuration. On the other hand, if the record is present, the ONU downloads the data. It is important to stress that because the ISPs may offer different transmission speeds, functions, etc., to customers, each customer should have his or her own distinct software image. The software image message responds to the image data transfer used to set the parameters. The message parameters are reported as "inactive" as they are in the initial phase of the file download. The next analyzed message informs about the software image being valid and active. As soon as the ONU has the software image, it is capable of transferring the customer service support data as well as the metadata. To support VoIP telephony, which is a QoS-demanding service, the ONU downloads an additional configuration containing information such as the type of codec, constant bit rate allocation, and T-CONT priority. The next step is to set the parameters for VoIP service. This service is a key service used for the highest priority end units. Their setting corresponds to the priority operation, i.e., T-CONT1, in which a fixed bandwidth must be assigned. In the case of most of the service providers, this value is set to 512 kbit/s. This speed

ID	Type		Entity
367	Get	TCI priority: 1	ONU
368	Get response	TCI priority: 1; result reason: command processed successfully; vendor id: ZTEG	ONU
370	Get	TCI priority: 1	Software image
371	Get response	Result reason: command processed successfully; version: V3R016C00S917T; is committed: uncommitted; is active: inactive; is valid: valid	Software image
373	Get response	Is committed: committed; is active: active; is valid: valid	Software image
378	Get	VOIP configuration state	VOIP config data
379	Get response	VOIP configuration state inactive: configuration retrieval has not been attempted	VOIP config data
381	Get response	Profile version: 00000000	VOIP config data
400	Get all alarms		ONU DATA
401	Get all alarms response	OMCI alarms received on ME—physical path termination point ETHERNET UNI, instance—257, LAN-LOS No carrier at the Ethernet UNI.	ONU DATA

Table 2.
OMCI channel details in the analyzed GPON network.

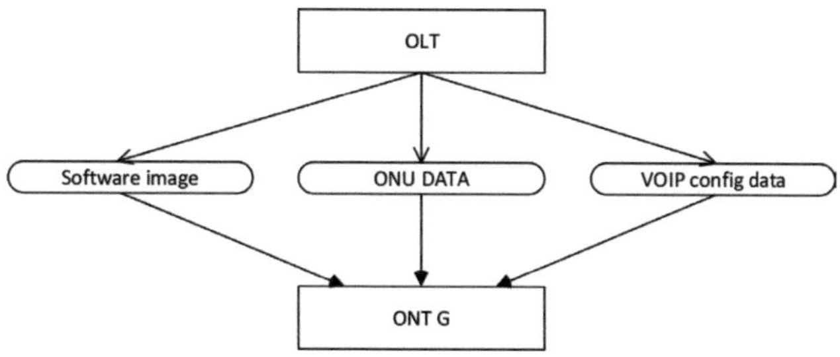

Figure 9.
OMCI channel diagram for the analyzed ONU.

must be guaranteed, even though it is considerably higher than the bandwidth of the G.711 codec (64 kbit/s). Successful reception and setting of the VoIP parameters are indicated with the message with ID 381 "Profile version: 00000000." In addition to the previously mentioned OMCI channel procedures that take place on the side of the ONU, there are also procedures on the side of the OLT: channel synchronization, verification, alarm indication, FEC monitoring, and so on, see **Table 2**. In summary, by analyzing the OMCI channel data, we performed active monitoring of the alarms of the distribution network. As seen in the "Get all alarms" message (ID: 400), the end-point ONU has reported a message signaling a failure on the Ethernet port.

In the case of OMCI channel measurement, it would be possible to summarize the transferred software image data, ONU data and the VoIP configuration file, see **Figure 9**.

A special case of the activation process is the message sequencing that can be seen in **Table 3**. This part of the activation is not mandatory for end units but is the last deactivation process aimed at the previously allocated parameters, most often the ONU-ID. This occurs when there is an immediate power outage. In the case of charged capacitors, the end unit sends a "Remote Error Indication" message. The message indicates that the ONU encountered an error. In the context of our experiments, this particular message was sent six times in total. When detecting a certain number of errors, most commonly defined by the manufacturer of the control unit, a "Dying Gasp" message follows. This message is dedicated to informing the control unit about an end unit failure, i.e., the loss of communication. The critical parameter of this message is the ONU-ID. After receiving such a message, the control unit sends the PLOAM message, "Deactivate ONU-ID," that causes this identifier to be released and consequently be reused by another end unit within the activation process. The PLOAM message is sent three times. Other parameters are discarded as internal timers have expired and communication/synchronization has not been restored in the downstream direction.

ID	ONU-ID	Message type	
1	1	Remote error indication	Sequence number: 3
6	1	Remote error indication	Sequence number: 8
7	1	Dying gasp	ONU ID: 1
9	1	Dying gasp	ONU ID: 1
59	1	Deactivate ONU-ID	Ident superframe counter: 498791449
60	1	Deactivate ONU-ID	Ident superframe counter: 498791451

Table 3.
Dying gasp PLOAM message details after the ONU lost power supply.

6. Conclusion

According to its grant policy, the European Union should contribute to building high-speed networks in the member states. This chapter introduced the state of the art in the field of Internet access technologies in the Czech Republic. The Czech Republic, as a member of the European Union, has committed to building high-speed Internet access for at least half of the households by 2020. Current market research has shown that WiFi technology is still dominant in the Czech Republic. The Czech Republic is behind the trend in FTTH/FTTB high-speed fiber optic connections by up to 10 and 5% for FTTB and FTTH, respectively.

The key part of this chapter is dedicated to the analysis of data transmitted in the GPON network. In cooperation with the Internet service provider Orange Slovakia, an active capture of transmitted data on the network was performed. As soon as the activation process of the end unit was completed successfully, data communication in both directions in GPON networks was possible. On the one hand, the sequence of the associated messages was defined by ITU-T Recommendation G.984, but on the other hand, it was only a recommendation and the specific implementation was fully within the manufacturer's competencies. Even though the end units were supposed to preserve the frame structure and the transmitted messages, as a result of the previously mentioned facts, it was often the case that the different manufacturers' end units were not compatible among themselves. Within the context of our analysis, TraceSpan's GPONxpert tool was used to capture network data. This device allowed for active listening of communication and real-time evaluation of its parameters. Detailed data analysis was a necessary form of postprocessing. To present the result of the activation process analysis, a sequence of key messages ensuring the activation of the end unit was displayed. Using these messages, it was possible to read the manufacturer and serial number of the end unit, set parameters such as ONU-ID and Alloc-ID. The OMCI channel provided end user parameters for a defined set of services, most often by downloading a profile image file corresponding to paid services and speeds. According to the reports, it was obvious that the VoIP parameters were also set.

Transmission of "Dying Gasp" messages was a special case of the activation process, or the logout and release of allocated parameters of the associated end units. These messages reflected a power outage of these units. Because the end units had unique UNU-ID/Alloc-ID parameters, the same parameters were used for other end units in the event of a power failure occurring in an already activated unit.

Acknowledgements

The presented research has been supported by a project of the Ministry of the Interior under grant no. VI20172019072, "E-infrastructure CESNET-modernization," registration no. CZ.02.1.01/0.0/0.0/16 013/0001797, and the National Sustainability Program under grant no. LO1401. For the research, the infrastructure of the SIX Center was used.

Conflict of interest

The authors declare no conflict of interest.

Notes/thanks/other declarations

Tomas Horvath would like to dedicate his part to his girlfriend (Lucie Baierova) and his family (Dagmar, Jan, and Petra). They have supported him during his University study. He also would like to give thanks to Ales Buksa for his support at the University. Ales has taught and inspired him with many things in his personal life.

Author details

Tomas Horvath[1,2]*, Petr Munster[1,2] and Josef Vojtech[2]

1 Department of Telecommunications, Brno University of Technology, Brno, Czech Republic

2 Department of Optical Networks, CESNET a.l.e., Prague, Czech Republic

*Address all correspondence to: horvath@feec.vutbr.cz

References

[1] Suzuki N, Miura H, Matsuda K, Matsumoto R, Motoshima K. 100 Gb/s to 1 Tb/s based coherent passive optical network technology. Journal of Lightwave Technology. 2018;**36**(8):1485-1491. DOI: 10.1109/JLT.2017.2785341

[2] Horvath T, Munster P, Vojtech J, Velc R, Oujezsky V. Simultaneous transmission of accurate time, stable frequency, data, and sensor system over one fiber with ITU 100 GHz grid. Optical Fiber Technology. 2018;**40**(1):139-143. DOI: 10.1016/j.yofte.2017.11.016

[3] Vojtech J, Slapak M, Skoda P, Radil J, Havlis O, Altmann M, et al. Joint accurate time and stable frequency distribution infrastructure sharing fiber footprint with research network. Optical Engineering. 2017;**56**(2):027101-027107. DOI: 10.1117/1.OE.56.2.027101

[4] G.983.1: Broadband Optical Access Systems Based on Passive Optical Networks (PON) [Internet]. International Telecommunication Union. Geneva: International Telecommunication Union; 1998. Available from: https://www.itu.int/rec/T-REC-G.983.1-199810-S/en [Accessed: 08-11-2018]

[5] Angelopoulos JD, Venieris IS, Protonotarios EN. A distributed FIFO spacer/multiplexer for access to tree APONs. In: Proceedings of ICC/SUPERCOMM'94-1994 International Conference on Communications. New Orleans, LA, USA: IEEE; 1994. pp. 70-74. Available from: http://ieeexplore.ieee.org/document/369020/

[6] 802.3ah-2004—IEEE Standard for Information technology—Local and metropolitan area networks—Part 3: CSMA/CD Access Method and Physical Layer Specifications Amendment: Media Access Control Parameters, Physical Layers, and Management Parameters for Subscriber Access Networks [Internet]. USA: IEEE; 2004. Available from: https://ieeexplore.ieee.org/document/983911 [Accessed: 08-11-2018]

[7] Kramer G, Pesavento G. Ethernet passive optical network (EPON): Building a next-generation optical access network. IEEE Communications Magazine. 2002;**40**(2):66-73. Available from: http://ieeexplore.ieee.org/document/983910/

[8] Next Generation Internet Initiative [Internet]. Next Generation Internet Initiative | Digital Single Market. EU; 2018. Available from: https://ec.europa.eu/digital-single-market/en/next-generation-internet-initiative [Accessed: 08-11-2018]

[9] G.984.1: Gigabit-Capable Passive Optical Networks (G-PON): General Characteristics [Internet]. International Telecommunication Union. Geneva: International Telecommunication Union; 2003. Available from: https://www.itu.int/rec/T-REC-G.984.1-200303-S/en [Accessed: 08-11-2018]

[10] Li Li, Xin Shouting, Duan De-Gong. Research of DBA schemes and QoS in PON system. In: 2016 2nd IEEE International Conference on Computer and Communications (ICCC). Chengdu, China: IEEE; 2016. pp. 2148-2153

[11] Arokkiam JA, Brown KN, Sreenan CJ. Optimised QoS-aware DBA mechanisms in XG-PON for upstream traffic in LTE backhaul. In: 2016 IEEE 4th International Conference on Future Internet of Things and Cloud Workshops (FiCloudW). Vienna, Austria: IEEE; 2016. pp. 361-368

[12] Horvath T, Munster P, Cymorek P, Oujezsky V, Vojtech J. Implementation of NG-PON2 transmission convergence

layer into OPNET modeler. In: 2017 International Workshop on Fiber Optics in Access Network (FOAN). Munich, Germany: IEEE; 2017. pp. 1-5

[13] Horvath T, Munster P, Vojtech J, Havlis O. Modified GIANT dynamic bandwidth allocation algorithm of NG-PON. Journal of Communications Software and Systems. 2017;**13**(1):15-22. DOI: 10.24138/jcomss.v13i1.243

[14] Farmer J, Lane BW, Bourg K, Wang W. FTTx Networks: Technology Implementation and Operation. Singapore: Morgan Kaufmann; 2017

[15] Digital Czech Republic v. 2.0—The Way to the Digital Economy [Internet]. Ministry of Industry and Trade. Prague: Ministry of Industry and Trade; 2014. Available from: https://goo.gl/J6ynRj [Accessed: 21-09-2018]

[16] National Plan for the Development of Next Generation Networks [Internet]. Ministry of Industry and Trade. Prague: Ministry of Industry and Trade; 2017. Available from: https://goo.gl/V3qxUD [Accessed: 21-09-2018]

[17] Y.2001: General Overview of NGN [Internet]. 2001: General Overview of NGN. Switzerland: ITU; 2005. Available from: https://www.itu.int/rec/T-REC-Y.2001/en [Accessed: 21-09-2018]

[18] Annual Reports 2014 [Internet]. Annual Rreports 2014. Prague: Czech Telecommunications Office (CTU); 2015. Available from: https://www.ctu.eu/2014-0 [Accessed: 21-09-2018]

[19] FTTx & Gigabit Society [Internet]. EU: IDATE; 2017. Available from: https://en.idate.org/categorie-produit/fttx-gigabit-en/ [Accessed: 30-09-2018]

[20] Montagne R. FTTH/B Panorama [Internet]. Spain; 2018. Available from: http://www.valencia.ftthconference.eu

[21] Rigby P. FTTH Handbook [Internet]. FTTHCouncil. EU: FTTHCouncil; 2016. Available from: http://www.ftthcouncil.eu/documents/Publications/FTTH_Handbook_V7.pdf [Accessed: 23-09-2018]

[22] Asaka K. What will be killer devices and components for NG-PON2?. In: 2014 The European Conference on Optical Communication (ECOC). Cannes, France: IEEE; 2014. pp. 1-3

[23] Nesset D. NG-PON2 technology and standards. Journal of Lightwave Technology. 2015;**33**(5):1136-1143. DOI: 10.1109/JLT.2015.2389115

[24] Khotimsky DA. NG-PON2 transmission convergence layer: A tutorial. Journal of Lightwave Technology. 2016;**34**(5):1424-1432. DOI: 10.1109/JLT.2016.2523343

[25] Horvath T, Munster P, Vojtech J, Havlis O, Gallo M. Transmission convergence layer of NG-PON2 in VPIphotonics tool. Journal of Communications Software and Systems. 2017;**13**(3):141-147. DOI: 10.24138/jcomss.v13i3.359

[26] G.989.3: 40-Gigabit-Capable Passive Optical Networks (NG-PON2): Transmission Convergence Layer Specification [Internet]. International Telecommunication Union. Geneva: International Telecommunication Union; 2015. Available from: http://www.itu.int/rec/T-REC-G.989.3/ [Accessed: 29-09-2018]

[27] Cale I, Salihovic A, Ivekovic M. Gigabit passive optical network—GPON. In: 2007 29th International Conference on Information Technology Interfaces. Cavtat, Croatia: IEEE; 2007. pp. 679-684

[28] Hood D, Trojer E. Gigabit-Capable Passive Optical Networks. Hoboken: Wiley; 2012

[29] G.984.3: Gigabit-Capable Passive Optical networks (GPON): Transmission Convergence Layer Specification. International Telecommunication Union. Geneva: International Telecommunication Union; 2015. Available from: http://www.itu.int/rec/T-REC-G.984.3/ [Accessed: 29-09-2018]

[30] Yan Y, Yamashita S, Yen S-H, Afshar PT, Gudla V, Kazovsky LG, et al. Invited paper: Challenges in next-generation optical access networks. IET Optoelectronics. 2011;**5**(4):133-143. DOI: 10.1049/iet-opt.2011.0027

[31] Mendonca C, Lima M, Teixeira A. Security issues due to reflection in PON physical medium. In: 2012 14th International Conference on Transparent Optical Networks (ICTON). UK: IEEE; 2012. pp. 1-4

[32] Horvath T, Munster P, Jurcik M, Koci L, Filka M. Timing measurement and simulation of activation process in GPON networks. Optica Applicata. 2015;**45**(4):459-470. DOI: 10.5277/oa150403

[33] Malina L, Munster P, Hajny J, Horvath T. Towards secure gigabit passive optical networks: Signal propagation based key establishment. In: Proceedings of SECRYPT 2015. Colmar, Francie: IEEE; 2015. pp. 349-354

[34] Malina L, Horvath T, Munster P, Hajny J. Security solution with signal propagation measurement for gigabit passive optical networks. Optik—International Journal for Light and Electron Optics. 2016;**127**(16):6715-6725. DOI: 10.1016/j.ijleo.2016.04.069

[35] Horvath T, Malina L, Munster P. On security in gigabit passive optical networks. In: 2015 International Workshop on Fiber Optics in Access Network (FOAN). Brno, Czech Republic: IEEE; 2015. pp. 51-55

Chapter 5

Research of M-PAM and Duobinary Modulation Formats for Use in High-Speed WDM-PON Systems

Toms Salgals, Inna Kurbatska, Sandis Spolitis, Vjaceslavs Bobrovs and Girts Ivanovs

Abstract

The exponential growth of Internet data traffic and progress of Information and Communication Technology (ICT) sector pushes hard the telecommunication infrastructure for upgrading the transmission data rate. Wavelength division multiplexed passive optical networks (WDM-PONs) can be the next generation solution for nowadays problems which are related to transmission capacity. Next-generation WDM-PON systems based on mixed wavelength transmitters are expected to become more cost-efficient at high per user data rates, e.g., over 10 Gbit/s per channel. Important advantage of this technology is to set various channel spacing and use different modulation formats to increase spectral efficiency in the same time and provide different transmission speeds for end user, based on pay-as-you-grow approach. Therefore, several modulation formats like non-return to zero (NRZ) also called 2-level pulse-amplitude modulation (PAM-2), four level PAM or PAM-4 and Duobinary (DB) are investigated to understand their limitations, advantages and disadvantages to be further used in next generation PON systems to increase its capacity and spectral efficiency.

Keywords: wavelength division multiplexed passive optical network (WDM-PON), non-return to zero (NRZ), four level pulse-amplitude modulation (PAM-4), duobinary (DB), capacity, spectral efficiency

1. Introduction

The exponential growth of Internet data traffic and progress of Information and Communication Technology (ICT) sector pushes hard the telecommunication infrastructure for upgrading the transmission data rate [1]. Power and cost-efficient fiber optical access networks, like passive optical network (PON) and short-range fiber optical links are one of the key technologies enabling bandwidth hungry services like video on demand (VoD), high definition TV, and cloud computing supported by large scale high-performance computers and data centers. Such optical links typically use direct detection and on-off keying modulation (OOK) with NRZ line code. Today's challenge for optical access networks and data centers is to increase the serial line rate of a NRZ link meeting the requirements to the physical bandwidth of the photonic and electronic components like optical signal modulators and photodiodes [2].

IntechOpen

Solution for telecommunication infrastructure upgrade and alternative solution for increase of the serial line rate of the NRZ link is to use multi-level signaling formats such as pulse-amplitude modulation (PAM), abbreviated as PAM-M or M-PAM, where multiple digital bits per symbol are encoded into M different signal amplitude levels. The four-level PAM modulation format is receiving significant attention because of its relative ease of implementation in comparison to higher-order modulation formats like quadrature phase-shift keying (QPSK), and m-ary quadrature amplitude modulation (m-QAM). It is clear that M-PAM offers a good trade-off between performance and complexity. Usage of PAM-4 format is effective way to double the data rate of NRZ link. Previously PAM-4 modulation formats have been investigated for application with traditional electrical networks [3, 4], but now researchers are focused on investigation of PAM-4 and M-PAM modulation formats for utilization in optical access networks as well as data center interconnections [5]. Also, there are very limited number of studies which are focused on spectrum slicing and stitching back method, which deals with bandwidth bottleneck problem by slicing the broadband signal in lower-bandwidth signal slices. This spectrum slicing and stitching back method or technique allows transmission of wide bandwidth signals from the service provider to the end user over an optical distribution network via low bandwidth equipment [6, 7]. It is ideally suited for cost sensitive fiber optical access networks where variable bandwidth and scalability as well as flexibility are important. It must be noted that this method is investigated for intensity modulated direct detection NRZ-OOK and duobinary systems, but there are no investigations on its usage together with M-PAM systems [8, 9]. It must be noted that multi-level signaling also changes some rules, which were used in NRZ coded transmission systems. For M-PAM systems it is important to implement more complex and precise level threshold detection for signal inputs, also signal-to-noise (SNR) requirements are higher than in case of NRZ. Eye time skew, amplitude compression in lower eye diagram eyes, intersymbol interference for M-PAM systems also is an issue which must be investigated. So, we can say that PAM-4 links are new science—still learning what impairments create errors in receivers [10, 11]. Significant efforts have been put on investigation of PAM-4 format in fiber optical transmission networks, however there are following aspects, which have not been studied or have been studied insufficiently. High-level PAM modulation techniques, like PAM-4, can dramatically improve the spectral efficiency and available bitrate by using the bandwidth of already existing optical, electro-optical or electrical devices. Minimal available channel spacing (which has direct impact on the utilization of resources like optical spectrum), maximal available number of channels, by wavelength division multiplexing (WDM) technique, maximal transmission distance (network reach) in dispersion compensated and non-compensated M-PAM modulated WDM-PON optical access systems.

Another way to improve capacity of limited bandwidth is by using duobinary modulation format. Transmission capacity will be increased in comparison with NRZ, utilization of DB will increase the transmission capacity by improving the bandwidth efficiency and reducing channel spacing with this modulation format [12]. Duobinary modulation format is type of proficient pseudo-multilevel modulation format, and therefore is the area of interest due to its increased spectral efficiency. It has been already used to increase the channel capacity by improving the bandwidth utilization in commercial links. The most important feature of duobinary modulation format is its usage for longer transmission distances where it has high tolerance to the influence of chromatic dispersion (CD) [13].

At first, in the paper we investigate the performance and minimal channel interval of 10 Gbit/s per channel NRZ-OOK (which is basically PAM-2) modulated transmission system, then we investigate PAM-4 and raise the transmission speed up to 20 Gbit/s per wavelength and in the end compare it to NRZ and duobinary modulation formats.

2. Evaluation of various channel spacings for increasing spectral efficiency of WDM-PON transmission system

At the moment passive optical networks have been standardized to next-generation NG-PON2 accordingly to ITU-T G.989.2 recommendation standards and are widely investigated. Operators are widely deploying time-division multiplexing (TDM) based passive optical networks in urban areas with bitrates up to 10 Gbit/s, but WDM-PON's still are in stage of research [14, 15].

The ITU-T G.694.1 recommendation provides a frequency grid for (WDM) transmission systems and specifies inter-channel intervals. The same frequency grid or channel spacing is used for spectral effectiveness improvement of PON system in our research. Anchored to 193.1 THz (central channel frequency), it supports a variety of inter-mediate channel spacings ranging from narrowed 12.5 GHz to 100 GHz and wider. Depending on the selected step of the inter-channel interval are defined the following abbreviations and acronyms:

WDM—wavelength division multiplexing.
CWDM—coarse wavelength division multiplexing.
DWDM—dense wavelength division multiplexing.

There are two types of inter-channel interval definitions in (WDM) systems:

Fixed inter-channel interval (fixed grid).
Flexible inter-channel interval (flexible grid).

According to ITU-T G.694.1 rec. the minimum step of a fixed channel interval is 12.5 GHz (please see **Table 1**). The flexible channel step is half of the 12.5 GHz, that can be used for the inter-channel interval like 6.25 GHz. Reducing the inter-channel interval leads to increase of crosstalk and non-linear effects (NOE) of transmitted optical signal [16–18].

For research of spectral efficiency increasing, the experimental 2-channel NRZ-OOK modulated 10 Gbit/s bit rate per channel transmission system model was created for Next-generation WDM-PON systems based on tunable wavelength transmitters, please see in **Figure 1**. First step of the research is based on various channel spacing impact on the end user transmitted signal with following fixed 10 Gbit/s transmission speed per channel.

As one can see in **Figure 1**. transmitter (Tx) part of our investigated transmission system model consists of two continuous wave (CW) laser sources—Agilent 81949A, with fixed central frequency 193.1 THz or 1552.524 nm in wavelength, and COBRITE DX-1 laser with tunable central frequency, which can be set the necessary channel spacing. Agilent 81949A continuous wave laser source was connected

Nominal central frequencies (THz) for spacing				Nominal central wavelengths (λ, nm)
12.5 GHz	25 GHz	50 GHz	100 GHz	
193.9375	-	-	-	1530.0413
195.9250	195.925	-	-	1530.1389
195.9125	-	-	-	1530.2365
195.9000	195.900	195.90	195.9	1530.3341

Table 1.
Nominal central frequencies grid of the DWDM grid [17].

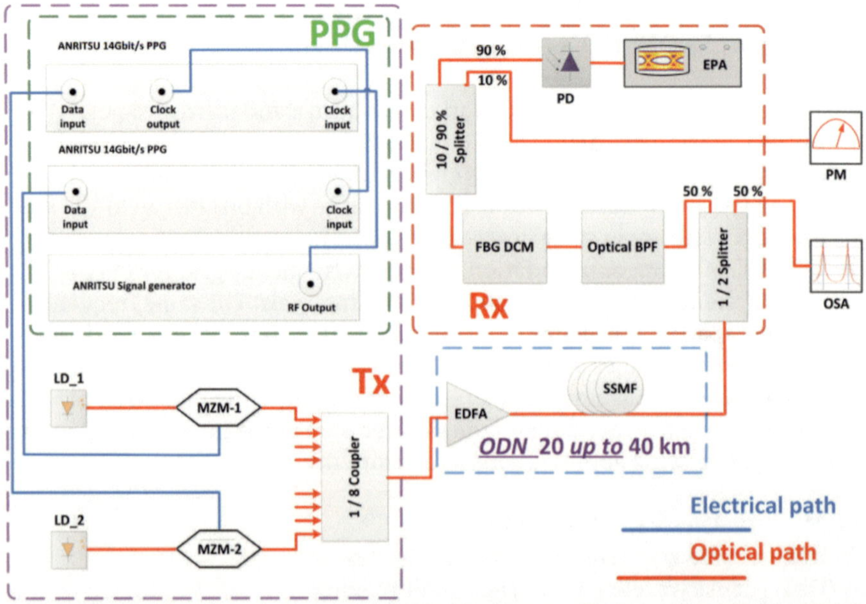

PPG: Pulse Pattern Generator, **Tx:** Transmitter, **Rx:** Receiver, **LD_1:** 1st CW Laser source, **LD_2:** 2nd CW Laser source, **MZM-1:** 10G MZM Mach-Zehnder-modulator, **MZM-2:** 40G MZM Mach-Zehnder-modulator, **EDFA:** EDFA amplifier, **SSMF:** ITU-T G.652 Single-mode-fiber, **OSA:** Optical Spectrum Analyzer, **Optical BPF:** Optical Tunable Filter, **DCM:** FBG Dispersion Compensation Module, **PD:** 10G Photodiode, **EPA:** Eye Pattern Analyzer.

Figure 1.
2-Channel NRZ-OOK modulated optical transmission system with 10 Gbit/s transmission speed per channel and flexible channel spacing.

to the 40G intensity Mach-Zehnder (MZM) modulator, COBRITE DX-1 laser light source was connected to the second MZM intensity modulator. Both laser sources were used with minimal output power +9 dBm for Agilent 81949A and +6 dBm for COBRITE DX-1. To provide the same level of optical power for both optical channels, after the PHOTLINE 40G MZM, an optical attenuator of 3.05 dB insertion loss was additionally attached to the modulator's optical output. Pulse Pattern Generator (PPG) with Pseudo random bit sequence (PRBS9) was used for generation of NRZ coded electrical signals. The external 10 GHz clock signal generator was used in this experiment for as a clock signal source for PPGs. Two electrical PPG non-inverted RF data signal outputs were connected to each of MZMs electrical signal inputs. The data rate for each of the PPGs was 10 Gbit/s throughout the experiment.

ITU-T G.652 standard single mode fiber (SSMF) with dispersion coefficient of 16 ps/(nm × km), and 0.2 dB/km attenuation coefficient was used in optical distribution network. Depending of SSMF fiber span length (20 or 40 km), an Erbium doped fiber amplifier (EDFA) with additional gain was used to provide sufficient optical power level before the PIN photoreceiver.

At the receiver part (Rx), the incoming optical signal was divided by 50% power splitter with 3.5 dB insertion loss. One output of optical power splitter was connected to the optical spectrum analyzer (OSA). Second output of power splitter was connected to the optical band pass filter (BPF) OTF-350 with a tuned 35 GHz 3-dB bandwidth. After BPF filter, fiber Bragg grating dispersion compensation module (FBG DCM), with 3 dB insertion loss was connected for post-compensation purposes of chromatic dispersion (CD). To avoid the maximum optical input optical power level rating of +3 dBm before the 10G PIN photoreceiver (PD) a monitoring power splitter with a power ratio of 10–90% and power meter was used. First channel was filtered out by using optical BPF. As one can see in **Figure 2(a)**, optical spectrum with central channel frequency 1552.560 nm (193.096 THz in frequency)

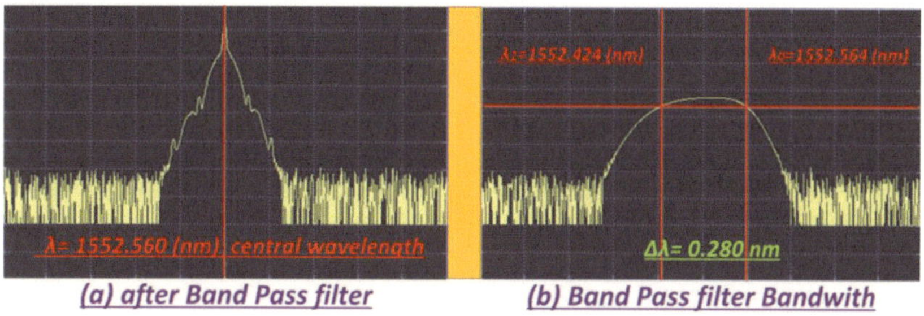

Figure 2.
Central channel spectrum of 2-channel NRZ-OOK modulated optical transmission system with 10 Gbit/s per channel: (a) after BPF and (b) measured amplited frequency response of BPF.

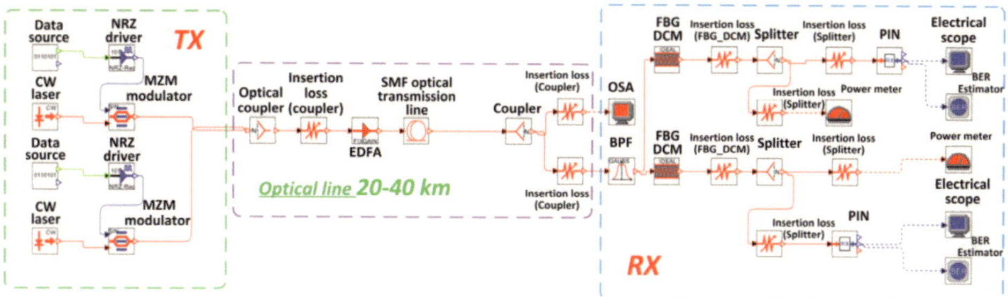

Figure 3.
Simulation scheme of 2-channel NRZ modulated optical transmission system with 10 Gbit/s transmission speed per channel with flexible channel interval.

is slightly shifted relative to ITU-T G.694.1 rec. Grid central frequency of 193.1 THz. By obtained results from the optical spectrum analyzer (OSA), the BPF pass band is $\Delta\lambda$ = 0.280 nm equal to 35 GHz, where λ_0 = 1552.564 nm and λ_1 = 1552.424 nm.

An eye analyzer was used for measurements of received electrical signal quality. The eyes of received signals for both channels were open, therefore leading to error free transmission. As the eye pattern analyzer for quality measurement use special masks to determine if the signal is above or below necessary quality. We continued our research in OptSim simulation environment by creating relevant simulation model and using the previously obtained experimental data.

For more precise expected Bit-error-rate (BER) values of received signal the simulation model was created in OptSim simulation software environment. The model used BER estimator based on statistical signal analysis. As one can see in **Figure 3**, simulation scheme implemented in OptSim simulation software for BER measurements has the same setup as experimental system. In the OptSim simulation environment, it is necessary to perform the assembly of used electrical-optical components in order to repeat the 2-channel NRZ-OOK modulated 10 Gbit/s per channel transmission system to research impact of various channel spacings.

According to ITU-T G.694.1 rec., see **Table 2**, during the experiment, the inter-channel interval for transmission system was changed from 100 GHz to 25 GHz. We started the experiment at a 20 km long fiber ODN distance with 100 GHz channel spacing. Firstly, the measurements was carried out without the chromatic dispersion (CD) post-compensation, at 20 km fiber link. For transmission over 20 km fiber span we observed negligible chromatic dispersion impact on 10 Gbit/s signal, received signal is mainly insignificant impact of dispersion [19].

The 12.5 GHz channel spacing interval was not obtained in this step of research. The reason for that was too wide filter pass-band, as a result photoreceiver captured

Frequency interval (THz)	**100 GHz**		**50 GHz**		**25 GHz**	
CW laser	Freq., (THz)	(λ, nm)	Freq., (THz)	(λ, nm)	Freq., (THz)	(λ, nm)
1st CH	193.1	1552.524	193.1	1552.524	193.1	1552.524
2nd CH	193.0	1553.328	193.05	1552.926	193.075	1552.725

Table 2.
Experimentally used channel interval according to ITU-T G.694.1 rec.

1st-CH, (THz)	**2nd-CH, (THz)**	**2nd-CH, (nm)**	**Delta, (THz)**	**CH–interval, (GHz)**
193.1	193.08125	1552.675	0.01875	18.750
193.1	193.07813	1552.700	0.02187	21.875
193.1	193.07500	1552.725	0.02500	25.000
193.1	193.07188	1552.751	0.02187	28.125
193.1	193.06875	1552.776	0.03125	31.250
193.1	193.06563	1552.801	0.02187	34.375
193.1	193.06250	1552.826	0.03750	37.500
193.1	193.05625	1552.876	0.04375	43.750
193.1	193.05000	1552.926	0.05000	50.000
193.1	193.00000	1553.329	0.10000	100.000

Table 3.
Channel spacing dependence on the channel interval.

both channels simultaneously. They did not appear on the Eye Analyzer because it was not possible to synchronize between the transmitter and receiver. After obtaining the results at fixed inter-channel intervals from 100 to 25 GHz, the smallest inter-channel interval at which transmission is possible was found. The step used to search for the inter-channel interval is 6.25 GHz and half of the found step 6.25/2 = 3.125 GHz. Result of channel spacing impact was obtained from channel with fixed central frequency of 193.1 THz = 1552.524 wavelength corresponding to the laser source used by Agilent 81949A. Our transmission system has only two channels, it is not possible to choose a central channel, both channels have mainly the same effect of crosstalk. The channel interval was changed by changing the central wavelength of the second CW laser source with 6.25 and 3.125 GHz step. Instead of experiment for 2-channel NRZ-OOK modulated optical transmission system with 10 Gbit/s transmission speed per channel previously calculated flexible channel interval was used in our research, please see **Table 3**.

Fiber optical transmission system made by the optical components affected by various factors caused by higher attenuation mentioned in specification insertion loss. To create same simulation model in OptSim simulation software environment, it was necessary to adapt model optical elements of the actual loss. In **Figure 4**. we can see BER estimated from the data obtained in OptSim simulation according to different channel intervals.

The BER threshold of 10^{-9} for our investigated transmission system was used to evaluate maximal crosstalk impact between the channels. According to the

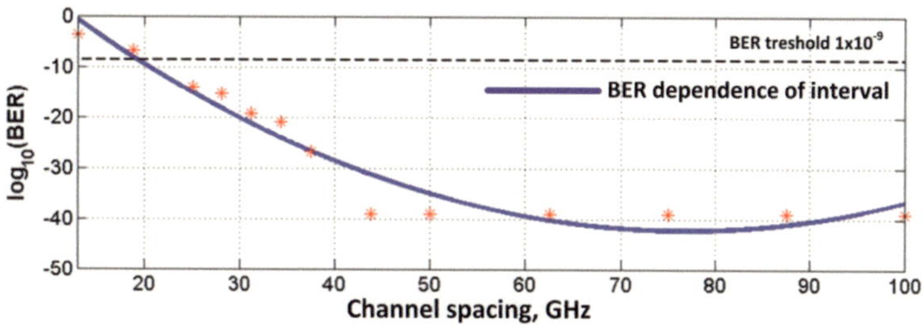

Figure 4.
BER dependence on the channel interval for a 20 km long 2-channel NRZ-OOK modulated optical transmission system with 10 Gbit/s transmission speed per channel.

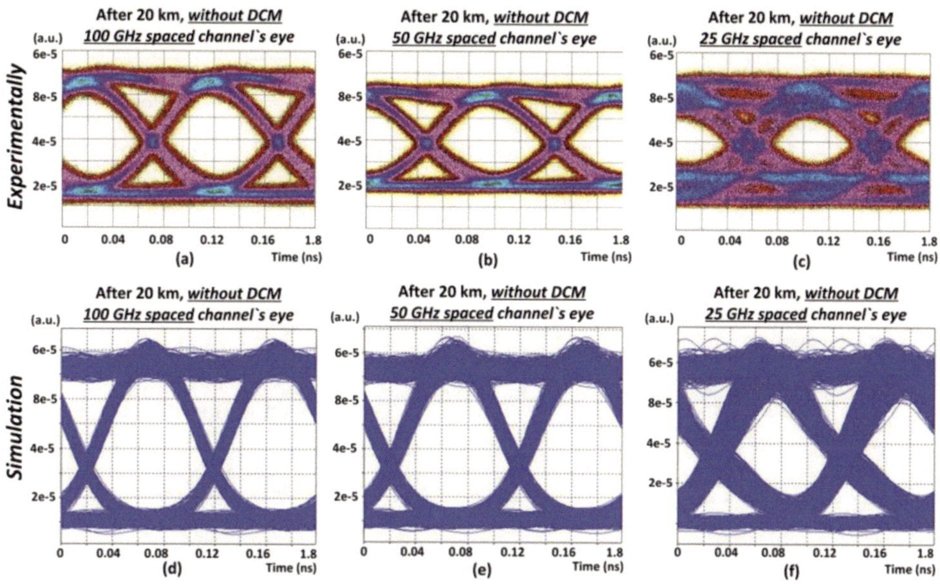

Figure 5.
Comparison of experimental and simulative results: eye diagrams of 20 km 2-channel NRZ-OOK modulated optical transmission system with 10 Gbit/s transmission speed per channel without CD post-compensation: (a) 100 GHz channel spacing, (b) 50 GHz channel spacing, (c) 25 GHz channel spacing, (d) 100 GHz channel spacing in the environment of OptSim, (e) 50 GHz channel spacing in the environment of OptSim, and (f) 25 GHz channel spacing in the environment of OptSim.

obtained results channel interval effect up to 30 GHz can be evaluated, higher than used value of BPF filter. Deterioration of the BER used for channel interval less than 30 GHz in our research, can be explained by adjacent channel overlapping. At 20 km long SSMF fiber optical link minimal channel spacing was achieved ensuring BER < 10^{-3} threshold at 25 GHz. In **Figure 5**, we can see experimental and theoretical (simulation data) eye diagrams of received signal for second channel with 100, 50 and 25 GHz channel spacing crosstalk impact, please see **Figure 5**.

In second part of our research the length of ODN was increased from 20 to 40 km, by adding 20 km SSMF fiber span. The effect of chromatic dispersion was observed in upgraded transmission system. Fiber Bragg grating dispersion compensation module (FBG DCM) with −640 ps/nm was used for dispersion compensation. The BER value exceeded our defined BER threshold of $1x10^{-9}$ at 31.25 GHz channel spacing according to the obtained results of OptSim simulation software. By performing experiment, the 31.25 GHz inter-channel spacing was the last interval at which mask testing with eye diagram analyzer for received eye diagrams was

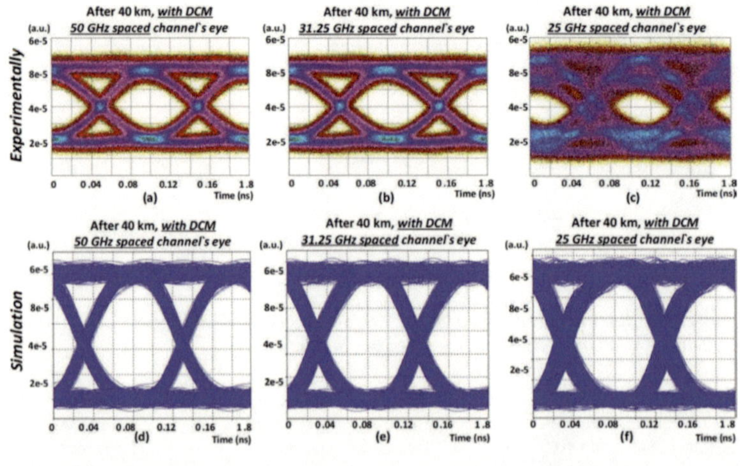

Figure 6.
Comparison of experimental and simulative results: eye diagrams of 40 km 2-channel NRZ modulated optical transmission system with 10 Gbit/s transmission speed per channel with CD post-compensation: (a) 50 GHz channel spacing, (b) 31.25 GHz channel spacing, (c) 25 GHz channel spacing, (d) 50 GHz channel spacing in the environment of OptSim, (e) 31.25 GHz channel spacing in the environment of OptSim, and (f) 25 GHz channel spacing in the environment of OptSim.

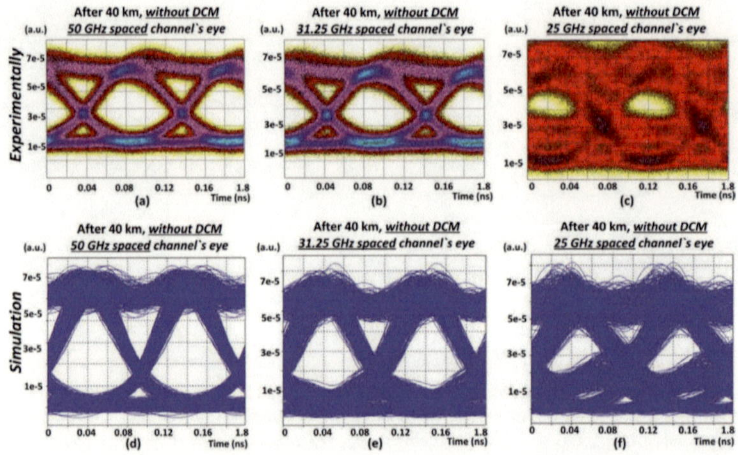

Figure 7.
Comparison of experimental and simulative results: eye diagrams of 40 km 2-channel NRZ modulated optical transmission system with 10 Gbit/s transmission speed per channel without CD post-compensation: (a) 50 GHz channel spacing, (b) 31.25 GHz channel spacing, (c) 25 GHz channel spacing, (d) 50 GHz channel spacing in the environment of OptSim, (e) 31.25 GHz channel spacing in the environment of OptSim, and (f) 25 GHz channel spacing in the environment of OptSim.

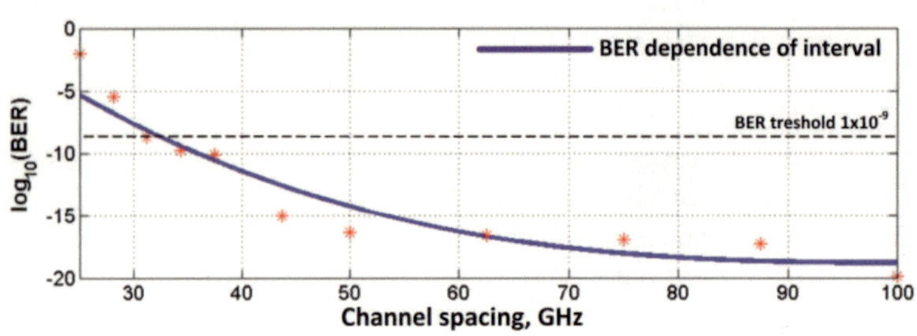

Figure 8.
BER dependence on channel interval for a 40 km 2-channel NRZ-OOK modulated optical transmission system with 10 Gbit/s transmission speed per channel.

possible [20]. By obtained experimental and simulation results it can be concluded that the model of optical transmission created in the simulation environment corresponds to the experimental fiber optic transmission system. Channel overlaps at 40 km long fiber section, with use of dispersion compensation, see **Figure 6** and without dispersion compensation see **Figure 7**. Results, with BER below our inter-channel interval, please see **Figure 8**.

Our defined BER threshold of 1×10^{-9} was exceeded at the 31.25 GHz channel interval where the BER of received signal was 7.4×10^{-11}.

3. Evaluation of PAM-4 modulation format use in WDM-PON systems

In our research we investigated the 4-channel 10 Gbaud/s (20 Gbit/s) per channel PAM-4modulated WDM-PON access system with minimal allowable channel spacing, which has a direct impact on the utilization of resources like optical spectrum. The research was made with and without fiber chromatic dispersion (CD) fiber Bragg grating compensation module (FBG DCM). We evaluate system performance and found the maximal transmission distance for multichannel PAM-4 modulated WDM-PON transmission system operating at 20 Gbit/s per channel. In OptSim simulation software we created transmission system model to evaluate the performance of 4-channel PAM-4 modulated WDM-PON transmission system operating at 10 Gbaud/s or 20 Gbit/s per channel under the condition with BER threshold of 10^{-3}, by use of Reed Solomon (RS 255,223) forward error correction (FEC) code for 10 Gbit/s PONs [21, 22]. The theoretical FEC relationship restores 1.1×10^{-3} pre-FEC BER to a 10^{-12} post-FEC in the PON standards. As it is shown in **Figure 9**, the PAM-4 modulated WDM-PON simulation scheme was created in OptSim simulation software environment. Here the Matlab software was used for BER estimation of received PAM-4 signals. WDM-PON simulation model consists of 4 channels, with central frequency 193.1 THz for second channel and chosen 50 or 100 GHz, according to the previously mentioned ITU G.694.1 rec. According to our previously channel interval research of flexible channel spacing like 37.5 and 25 GHz also was realized. However, the quality of received signal was low, with crosstalk impact and error-free transmission was not possible, performance was above our defined BER threshold 1×10^{-3}.

We evaluated the performance of WDM-PON architecture in terms of maximal transmission reach. Optical line terminal (OLT) is located in central office (CO) and consists of four transmitters (OLT_Tx). Each OLT_Tx transmitter consists of two pseudo-random bit sequence (PRBS) generators and NRZ drivers, as a result two

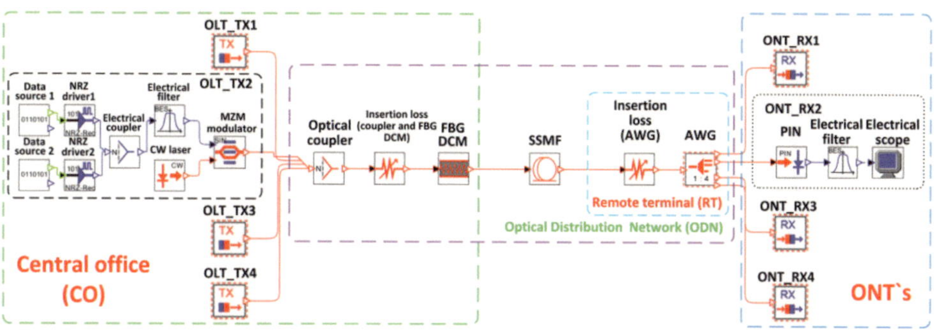

Figure 9.
Simulation scheme of 4-channel PAM-4 modulated WDM-PON transmission system operating at 10 Gbaud/s per wavelength.

electrical signals are generated where one of them has twice larger amplitude than other for each particular electrical signal. An electrical coupler is used to couple both electrical signals in such a way generating electrical PAM-4 signal. Afterwards, additional electrical filters were used for ensuring of optimal system performance. Generated PAM-4 signal was send to external MZM with 3 dB insertion loss and 20 dB extinction ratio. Continuous wavelength (CW) laser with linewidth of 50 MHz and output power of +3 dBm is used as the light source [23].

Optical signals from four transmitters are coupled together by using optical coupler with 1 dB insertion loss. Chromatic dispersion pre-compensation by FBG DCM, with additional 3 dB insertion loss is realized for all channels before launching optical signal in ITU-T G.652 single mode fiber (SMF), used for transmission in optical distribution network (ODN). After transmission in ODN, all channels are separated by arrayed waveguide grating (AWG) demultiplexer which insertion loss is 3.5 dB. Here we applied various channel spacings—50 or 100 GHz (3-dB bandwidth is 20 GHz) for research of the crosstalk impact. Each receiver of optical network terminal (ONT) consists of PIN photoreceiver (sensitivity is −19 dBm for BER of 10^{-12}). An optimal electrical Bessel low-pass filter (LPF) with bandwidth (3-dB bandwidth is 7.5 GHz), was adopted for more successful system performance. An electrical scope was used for evaluation of received signal bit patterns quality, accordingly, eye diagrams.

As it is shown in **Figure 10(a)** in B2B configuration for first investigated 100 GHz channel spacing, the signal quality is good, eye is open and error-free transmission can be provided. After 59 km transmission which was the maximum transmission distance without use of FBG DCM, the BER of received signal was 7.5×10^{-4}, please see **Figure 10(b)**. Dispersion compensation FBG DCM module was implemented to evaluate transmission distance in terms of maximal reach. As it is shown in **Figure 10(c)** by using this technique of FBG DCM, the maximum achievable transmission distance 74 km was reached, where BER of received signal was 9×10^{-4}. Extra 15 km or 25.4% of link length was gained.

Therefore, basis on our research data we can conclude that narrower channel spacing for 4-channel PAM-4 10 Gbaud/s WDM-PON system is 50 GHz. As it is shown in **Figure 11(a)** in B2B configuration for second investigated 50 GHz channel spacing, the signal quality is good, eye is open and error-free transmission can be provided. After 58 km transmission, which was the maximum transmission distance without use of FBG DCM, the BER of received signal was 8×10^{-4}, shown in **Figure 11(b)**. In our research we show the eye diagrams of received signal for the second channel, the drop in BER performance can be explained by the impact of crosstalk between channels. Dispersion compensation FBG DCM module was

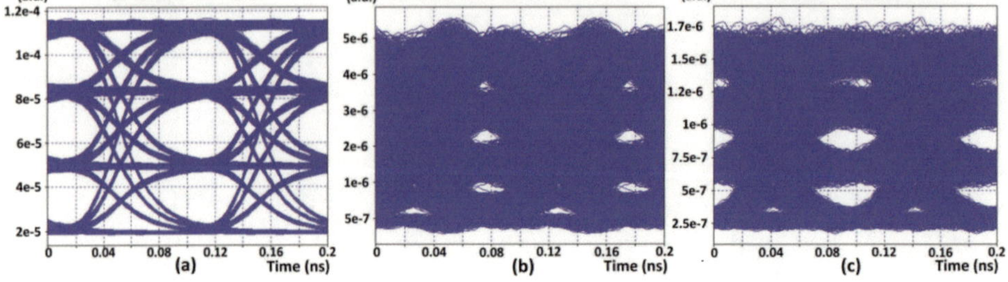

Figure 10.
Eye diagrams of received signal (a) after B2B transmission, (b) after 59 km transmission without use of CD pre-compensation, (c) after 74 km transmission with use of CD pre-compensation for 4-channel 20 Gbit/s per channel PAM-4100 GHz spaced WDM-PON transmission system.

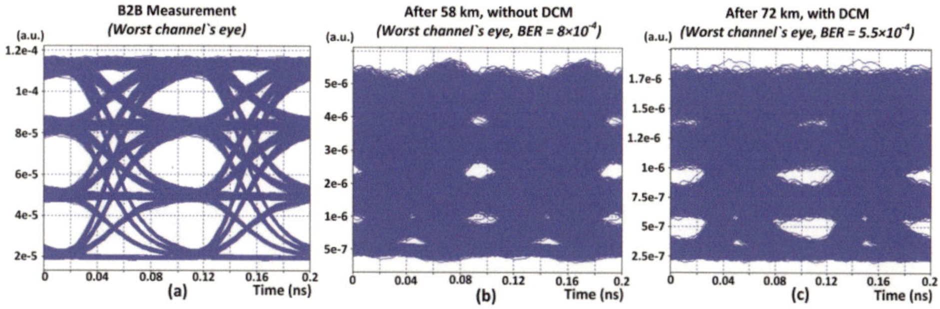

Figure 11.
Eye diagrams of received signal (a) after B2B transmission, (b) after 58 km transmission without use of CD pre-compensation (c) after 72 km transmission with use of CD pre-compensation for 4-channel 20 Gbit/s per channel PAM-4 50 GHz spaced WDM-PON transmission system.

implemented to evaluate transmission distance in terms of maximal reach. As it is shown in **Figure 11(c)**, by using this technique of FBG DCM, the maximum achievable transmission distance was 72 km, with BER of received signal 5.5×10^{-4}. Extra 14 km or 24% of link length was gained.

It was shown, that maximal transmission distance with BER below FEC limit of 10^{-3} for 100 GHz spaced 4-channel PAM-4 WDM-PON system can be increased by 15 km or 25.4% by use of implemented FBG DCM. In case of 50 GHz channel spacing, maximum transmission system reach can be increased by 14 km or 24% by use of FBG DCM.

4. Evaluation of PAM-4, NRZ and duobinary modulation formats performance in WDM-PON system architecture

In case of research we improve our previously made 4-channel PAM-4 WDM-PON system simulation model capacity by increasing number of multilevel channels and implement the use of different modulation formats in terms of system performance by maximal achievable reach. Several modulation formats have been proposed in the past and have become standards. In this research are investigated several modulation formats for use in WDM-PON architecture-based system, like NRZ, PAM-4 and duobinary (DB). Alternative solution instead widely used direct detection on-off keying modulation format NRZ-OOK with physical bandwidth limitations is to use more spectrally efficient multi-level formats such as PAM-4 [24, 25]. Another way to improve the bandwidth efficiency and reduce channel spacing is by using duobinary modulation format [12]. The most important feature of this multi-level modulation format duobinary is a viability of usage for longer transmission distances without regeneration with high tolerance to chromatic dispersion CD influence. As we know duobinary is used to increase the channel capacity by improving the bandwidth utilization [13].

The goal of our created 8-channel 20 Gbit/s per channel WDM-PON simulation model evaluate maximum transmission reach using different modulation formats, discussed previously in this paper like NRZ, PAM-4 and perspective duobinary modulation format. As it is shown in **Figure 12** the 8-channel WDM-PON simulation scheme with different optical transmitters (Tx) located in CO Optical Line Terminal (OLT_Tx) part for each modulation format realization are shown. According to ITU-T G.694.1 rec. Frequency with grid central frequency of 193.1 THz and channel spacing of 50 and 100 GHz are chosen for research of crosstalk impact on modulation formats under research [26].

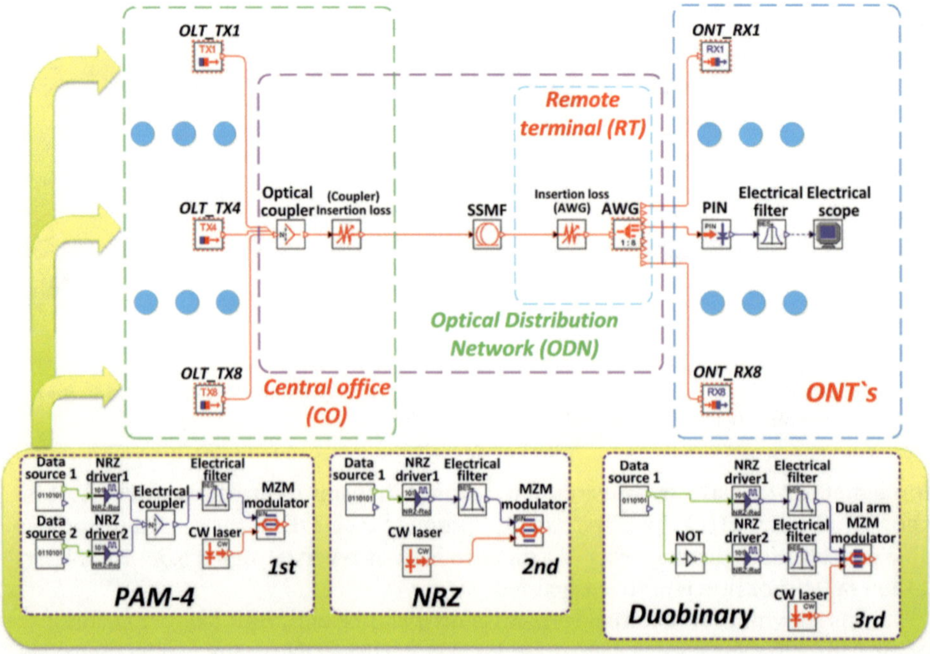

Figure 12.
Simulation scheme of 8-channel 20 Gbit/s transmission speed per channel PAM-4, DB and NRZ modulated WDM-PON optical transmission system.

In first simulation model PAM-4 transmitter is designed like previously, from two 10 Gbit/s NRZ coded electrical data signals (where one of them has twice larger amplitude), by coupled together with electrical coupler. Coupled PAM-4 electrical signal filtered with electrical Bessel low-pass filter (3-dB bandwidth is 10 GHz) and send to external MZM [21].

Second simulation model duobinary transmitter was realized with 20 Gbit/s bit rate per channel. Data source element with pseudo random bit sequence (PRBS) has only one logical output, where the output signal is divided in two signals. One of those signals is inverted by logical NOT element. Afterwards each data signal sent to NRZ drivers and filtered by Bessel low-pass filters (3-dB bandwidth is 5 GHz). Each NRZ coded electrical signal is passed to inputs of dual-arm MZM, at the end forming the DB transmitter [27].

Third simulation model NRZ transmitter consists of one NRZ driver with electrical signal input of data source with PRBS sequence. Afterwards NRZ coded data signal are directly connected to MZM RF signal input.

Following fixed parameters of optical and electrical elements was used: continuous wavelength (CW) laser output power + 6 dBm, extinction ratio 20 dB and 3 dB insertion loss of MZM, ITU-T G.652 SSMF with dispersion coefficient 17 ps/(nm × km), dispersion slope 0.056 ps/nm^2 × km and 0.2 dB/km attenuation coefficient [28]. Bandwidth of electrical LPF filters has been adjusted for optimal performance of each modulation format and have not been changed during research.

Each receiver consists of 40 GHz PIN photodiode with sensitivity equal to −19 dBm at 10 Gbit/s reference bit rate, dark current of 10 nA and responsivity of 0.8 A/W [29]. An electrical LPF filter bandwidth was adopted at receiver side for more successful system performance depending on the used modulation format. During the simulations LPF bandwidth of 15 GHz was chosen for PAM-4 modulated signals, and 10 and 17 GHz for DB and NRZ modulated electrical signals.

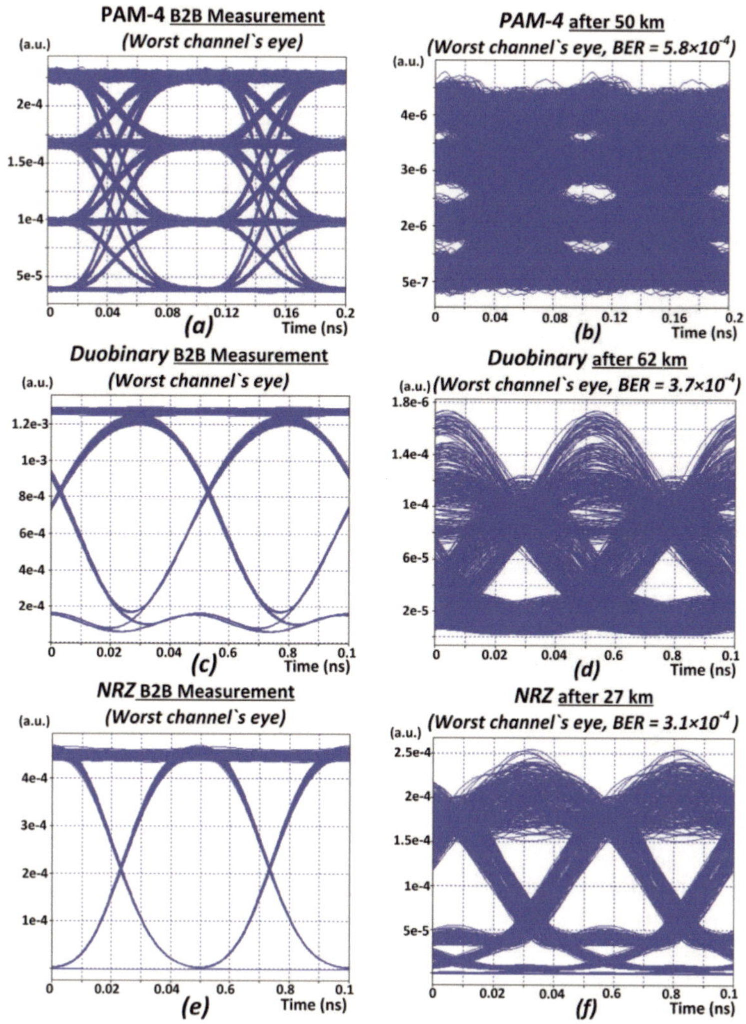

Figure 13.
Eye diagrams of received (a) PAM4, (c) DB and (e) NRZ signals after B2B transmission, and after maximal reached transmission distance: (b) 50 km with PAM-4, (d) 62 km with DB, (f) 27 km with NRZ modulated signals for 8-channel 20 Gbit/s per channel WDM -PON transmission system.

The BER threshold of 10^{-3} with additional FEC was used for our investigated WDM-PON transmission system to compare performance in terms of maximal network reach for PAM-4, DB, NRZ modulated optical signals. During the simulations it was observed that maximal achievable distance has minimal crosstalk impact on BER for all modulation formats, which was negligible, depending on our chosen channel spacing.

As it is shown in **Figure 13(a, c and e)** in B2B configuration for narrowest investigated 50 GHz channel spacing, the signal quality is good, eye is open and error-free transmission can be provided. After transmission the BER of received DB modulated signal with maximum reached distance of 62 km was 3.7×10^{-4}. PAM-4 and NRZ modulated signals shows 50 km and 27 km maximal reached transmission distance, where BER of received signal was 5.8×10^{-4} and 3.1×10^{-4}, please see **Figure 13(b, d and f)**. The largest network reach with BER below defined threshold, was provided by DB modulation format, extending the reach of 62 km.

5. Conclusions

Nowadays the WDM-PON systems rely on fixed wavelength transmitters and are expected to become more cost-efficient at high per user data rates. It was examined that different types of optical modulation formats are available for passive WDM fiber optical access networks. Implementation and research of multilevel modulation formats like PAM-4 and duobinary can dramatically improve the spectral efficiency and available bitrate by using the bandwidth of already existing optical, electro-optical or electrical devices. Theoretical simulations and experimental research methods showed possibility to double the available transmission speed in optical access networks by using the same bandwidth, e.g., instead of 10 Gbit/s transmit 20 Gbit/s signal by using 10 GHz electrical and electro-optical equipment, if PAM-4 modulation method is used. In our research we investigated existing optical modulation formats—widely used NRZ, DB and PAM-4 for optical access networks, by experimentally demonstrating and modeling system transmission in RSOFT OptSim simulation environment and Matlab software. As it shown by simulation results, narrowest channel spacing provides higher spectral efficiency. However, better signal quality and system performance are achieved with larger channel spacing interval, e.g., 100 GHz, mainly due to crosstalk between channels. From experimental data we can clearly see that the chromatic dispersion limits transmission capacity when bit rates increase. Implementation of the efficient compensation solution may sufficiently extend the reach of optical link and improve the transmission quality in our investigated WDM-PON systems.

Acknowledgements

This work has been supported by the European Regional Development Fund within the Activity 1.1.1.2 "Post-doctoral Research Aid" of the Specific Aid Objective 1.1.1 "To increase the research and innovative capacity of scientific institutions of Latvia and the ability to attract external financing, investing in human resources and infrastructure" of the Operational Programme "Growth and Employment" (No. 1.1.1.2/VIAA/1/16/044).

Author details

Toms Salgals*, Inna Kurbatska, Sandis Spolitis, Vjaceslavs Bobrovs and Girts Ivanovs
Institute of Telecommunications, Riga Technical University, Riga, Latvia

*Address all correspondence to: toms.salgals@rtu.lv

References

[1] Wei JL, Grobe K, Griesser H. High speed next generation passive optical networks: Performance, cost, and power dissipation. In: 2016 Progress in Electromagnetic Research Symposium (PIERS), Shanghai, China. 2016. pp. 4856-4857

[2] Reed GT, Mashanovich G, Gardes FY, Thomson DJ. Silicon optical modulators. Nature Photonics. 2010;**4**(8):518-526

[3] Nazemi A, Kangmin H, Catli B, Delong C, Singh U, He T, et al. A 36Gb/s PAM4 transmitter using an 8b 18GS/s DAC in 28 nm CMOS. In: ISSCC 2005 IEEE International Digest of Technical Papers Solid-State Circuits Conference. 2015. pp. 1-3

[4] Amirkhany A, Abbasfar A, Savoj J, Jeeradit M, Garlepp B, Kollipara RT, et al. A 24 Gb/s software programmable analog multi-tone transmitter. IEEE Journal of Solid-State Circuits. 2008;**43**(4):999-1009

[5] Xiong C, Douglas MG, Jonathan EP, Jason SO, Wilfried H, William MJG. Monolithic 56 Gb/s silicon photonic pulse-amplitude modulation transmitter. Optica. 2016;**3**:1060-1065

[6] Geisler DJ, Fontaine NK, Scott RP, He T, Paraschis L, Gerstel O, et al. Bandwidth scalable, coherent transmitter based on parallel synthesis of multiple spectral slices. In: Optical Fiber Communication Conference and Exposition (OFC/NFOEC), 2011 and the National Fiber Optic Engineers Conference. Los Angeles, CA; 2011, paper OTuE3. pp. 1-3

[7] Spolitis S, Olmos JJ, Bobrovs V, Ivanovs G, Monroy I. A novel approach for transmission of 56 Gbit/s NRZ signal in access network using spectrum slicing technique. In: Asia Communications and Photonics Conference 2013: Conference Proceedings, China, Beijing. 12-15 December 2013, paper AF4D.2. pp. 1-3

[8] Wagner C, Spolitis S, Olmos J, Bobrovs V, Tafur MI. Re-use of low bandwidth equipment for high bit rate transmission using signal slicing technique. In: Conference proceedings of Asia Communications and Photonics Conference, China, Hong Kong, 19-23 November 2015. Honkonga: Optical Society of America; 2015. pp. 1-3

[9] Spolitis S, Wagner C, Olmos J, Bobrovs V, Ivanovs G, Monroy I. Experimental Demonstration of a Scalable Sliceable Transceiver for Optical Access Networks. In: Asia Communications and Photonics Conference 2014: Conference Proceedings, China, Shanghai, 11-14 November 2014. Shanghai: Optical Society of America; 2014. pp. 1-3

[10] Keysight Technologies, Canada Keysight High Speed Digital Seminar Series. PAM-4 Stressed Pattern Generation. 2015. p. 41. Available from: http://www.keysight.com/upload/cmc_upload/All/HSDSeminarPaper4.pdf

[11] Keysight Technologies. PAM-4 Design Challenges and the Implications on Test. Application Note. 2015. p. 12. Available from: http://literature.cdn.keysight.com/litweb/pdf/5992-0527EN.pdf

[12] Kaur R, Dewra S. Duobinary modulation format for optical system—A review. International Journal of Advanced Research in Electrical, Electronics and Instrumentation Engineering. 2014;**3**:11039-11046. Available from: http://www.ijareeie.com/upload/2014/august/12_%20Duobinary.pdf

[13] Qamar F, Khawar Islam M, Shah SZA, Farhan R, Ali M. Secure duobinary signal transmission in optical communication networks for high performance & reliability. IEEE Access. 2017;**5**:17795-17802

[14] Kani J, Bourgart F, Cui A, Rafel A, Campbell M, Davey R, et al. Next-generation PON-part I: Technology roadmap and general requirements. IEEE Communications Magazine. 2009;**47**(11):43-49

[15] Sanchez RJ, Hernandez A, Larrabeiti D. Troubleshooting PON networks effectively with carrier-grade ethernet and WDM-PON. IEEE Communications Magazine. 2014;**52**(2):7-13. Available from: https://ieeexplore-ieee-org.resursi.rtu.lv/stamp/stamp.jsp?tp=&arnumber=6736739

[16] Bobrovs V. Analyzing and evaluation of channel interval in wavelength division multiplexing transmission systems [PhD thesis]. Riga: Riga Technical University; 2010. pp. 17-18

[17] ITU-T G.694.1 Spectral Grids for WDM Applications: DWDM Frequency Grid. 2012. pp. 1-7

[18] Olonkins S, Spolitis S, Lyashuk I, Bobrovs V. Cost effective WDM-AON with multicarrier source based on dual-pump FOPA. In: International Congress on Ultra Modern Telecommunications and Control Systems and Workshops, 2015. 2015. Art. No. 7002073. pp. 23-28

[19] Spolitis S, Bobrovs V, Ivanovs G. Reach improvement of spectrum-sliced dense WDM-PON system. In: Proceedings—2012 7th International Conference on Broadband, Wireless Computing, Communication and Applications, BWCCA 2012. 2012. pp. 296-301. Art. No. 6363072

[20] Anritsu. Enabling Precision EYE Pattern Analysis Extinction Ratio, Jitter, Mask Margin. pp. 3-6. Available from: https://dl.cdn-anritsu.com/en-us/test-measurement/files/Technical-Notes/Technical-Note/MP2100A_EE1100.pdf

[21] Brunina D, Porto S, Jain A, Lai CP, Antony C, Pavarelli N, et al. Analysis of forward error correction in the upstream channel of 10Gb/s optically amplified TDM-PONs. In: 2015 Optical Fiber Communications Conference and Exhibition (OFC), Los Angeles, CA. 2015. pp. 1-3

[22] ITU-T Recommendation G.984.2, Gigabit-capable Passive Optical Networks (GPON): Physical Media Dependent (PMD) layer specification. 2011, 1-36. Available from: https://www.google.com/url?sa=t&rct=j&q=&esrc=s&source=web&cd=2&ved=2ahUKEwigteS7mMXgAhVl6KYKHU1UD6EQFjABegQICRAC&url=https%3A%2F%2Fwww.itu.int%2Frec%2Fdologin_pub.asp%3Flang%3De%26id%3DT-REC-G.984.2-200303-I!!PDF-E%26type%3Ditems&usg=AOvVaw2nWcdEiNYzk3CnHZJ8tPgR

[23] Salgals T, Spolitis S, Olonkins S, Bobrovs V. Investigation of 4-PAM modulation format for use in WDM-PON optical access systems. In: Progress in Electromagnetics Research Symposium (PIERS 2017), St. Petersburg, Russia. 2017. pp. 1-4

[24] Forzati M, Berntson A, Martensson J, Pincemin E, Gavignet P. NRZ-OOK transmission of 16 × 40 Gb/s over 2800 km SSMF using asynchronous phase modulation. In: IEEE Conference on Lasers and Electro-Optics and 2008 Conference on Quantum Electronics and Laser Science. 2008. pp. 1-2

[25] Wang Y, Gai W, Tang L. A novel 40-Gb/S PAM4 transmitter with power-efficient pre-emphasis. In: IEEE International Conference on Solid-State and Integrated Circuit Technology (ICSICT). 2014. pp. 1-3

[26] Salgals T, Skladova L, Vilcane K, Braunfelds J, Spolitis S. Evaluation of 4-PAM, NRZ and duobinary modulation formats performance for use in 20 Gbit/s DWDM-PON optical access systems. In: 2018 Advances in Wireless and Optical Communications

(RTUWO 2018): Proceedings, Latvia, Riga. 15-16 November 2018

[27] Kurbatska I, Spolitis S, Bobrovs V, Alševska A, Ivanovs Ģ. Performance comparison of modulation formats for 10 Gbit/s WDM-PON systems. In: 2016 Advances in Wireless and Optical Communications (RTUWO 2016): Proceedings, Latvia, Riga, 3-4 November 2016. Piscataway: IEEE; 2016. pp. 51-54

[28] ID Photonics GmbH. CoBrite DX1–Laser. Technical Specification. 2017. pp. 1-4. Available from: https://www.id-photonics.com/images/stories/PDF/Data_sheet_CBDX1-x-x-xx.pdf

[29] Optilab Technical Specification of "40 GHz Linear InGaAs PIN Photodetector". 2016. pp. 1-4. Available from: https://www.oequest.com/getDatasheet/id/9716-9716.pdf

Chapter 6

Mitigating Turbulence-Induced Fading in Coherent FSO Links: An Adaptive Space-Time Code Approach

Ojo O. Adedayo, Oluwafemi B. Ilesanmi, Ogunlade M. Adegoke and Ajibade Adedayo

Abstract

Free space optical communication systems have witnessed a significant rise in attention over the last half a decade owing largely to their enormous bandwidth and relative ease of deployment. Generally, free space optical communication systems differ in their detection mechanism as various detection mechanisms are being reported, including intensity modulation/direct detection FSO, differential FSO and coherent FSO. In this chapter, we explore the prospect of obtaining an optimally performing FSO system by harnessing the cutting-edge features of coherent FSO systems and the coding gain and diversity advantage offered by a four-state space-time trellis code (STTC) in order to combat turbulence-induced fading which has thus far beleaguered the performance of FSO systems. The initial outcomes of this technique are promising as a model for various visible light communication applications.

Keywords: free space optical communication, space-time trellis code, turbulence, coherent detection

1. Introduction

Telecommunication is one of the most important innovations in the history of mankind as it affords people the opportunity to communicate rapidly and reliably over long distances, often breaking physical and geographical barriers to make the world a global village as it is known today. One of the key ingredients in the heart of communication technologies over the last century is wireless communication. The advent of RF wireless communication techniques and protocols has been instrumental to the giant strides made in the communication domain as it eliminates the cumbrous requirement of lengthy wired connections, a requirement which has often been a great limitation for wired communication systems.

RF wireless communication systems enjoyed significant attention and penetration but soon became a victim of its success as more and more contents are demanded by users due to the proliferation of data, video, gaming and general broadband multimedia. These demands have prompted communication system engineers to explore more efficient, faster and reliable wireless communication techniques. To this end, free space optical (FSO) communication has been a viable solution [1].

Free space optical communication is a communication technology that employs light as carrier by modulating baseband information with optical carriers often from laser beams through free space to the receiver [2]. The path of connection between FSO transmitters and the receivers are known as FSO links. Even though the very first optical system dates back to the eighteenth century, modern FSO communication systems were first widely deployed by the National Aeronautics and Space Administration (NASA) and have since become a promising broadband wireless access technology. FSO communication systems are also being combined with standard RF systems in order to form hybrid communication systems that harness the unique features of RF and FSO communication systems to enhance performance, capacity and reliability.

Characteristically, FSO communication systems are highly secure as they have high immunity to interference with the use of secure point-to-point line-of-sight links, they require no licensing or regulatory permission, and they are fast and can be easily deployed and operated compared to other systems like the fiber optic systems [3]. These features make FSO communication the favored option for the provision of high-speed links for a variety of next generation optical applications including broadcast, security, wireless backhaul at a data rate as high as 40 Gbps [4], fiber backup and last mile communication [5]. Finally, the ease of setup and cost effectiveness of FSO systems have made them the preferred option for restoring connection in case of disaster.

However, the performance of FSO communication systems are greatly affected by turbulence-induced fading [6–12], and different investigations are currently being explored to address this challenge. The inhomogeneity of the temperature and pressure of the atmosphere causes local variations in the refractive index as light propagates from the transmitter to the photoreceptor; these variations degrade the performance of FSO links significantly.

Geared towards the improvement of the performance of coherent FSO communication systems in the presence of atmospheric turbulence, this work examines the error reduction schemes currently being employed for FSO links and presents an adaptive space-time trellis code (STTC) scheme for coherent FSO links.

2. Free space optical communication: types and variants

In terms of reception technique, however, the most commonly reported variants of FSO communication systems are the direct detection/intensity modulated (IM/DD) FSO system and the coherent FSO system. The IM/DD FSO communication systems convey the information to be transmitted only on the intensity of the emitted light, and the receivers simply decode the information as the light changes in intensity. In coherent FSO communication systems, however, other signal properties such as phase and frequency may be employed in conveying the information. At the receiving end, as against simply observing changes in light intensity as in the case of IM/DD FSO systems, coherent FSO communication systems, first, mix the received field optically with a local oscillator before the actual photodetection. So far, more works on the IM/DD FSO communication system are being reported in literature owing to its simplicity of detection as less complex receivers and algorithms are required. Coherent FSO systems, though more complex, however, offer superior performance in terms of improved receiver sensitivity and background noise rejection [13].

Finally, in terms of communication range, FSO communication systems can be characterized into short-range, medium-range, long-range and inter-terrestrial FSO systems depending on the application for which they are deployed, and these applications include inter-chip communication, inter-vehicular communication, metropolitan area communication as well as satellite and space exploration.

3. FSO system and channel models

Concerted efforts being expended by researchers in the quest of effectively modeling the FSO channel are geared towards understanding the channel and serving as the template upon which FSO modulators, demodulators, receivers and other devices can work. Accurate mathematical models for FSO communication system are the basis upon which the development of high-performing hardware is established, despite huge technical challenge of turbulence-induced fading. This challenge is however being addressed using various techniques as summarized in **Figure** 1.

In the wavelength diversity schemes, the source information is encoded into different wavelengths obtainable from different constituents of the infrared spectrum.

4. Turbulence models

One of the most crucial steps in the attempts to mitigate the degradation in the performance of optical communication systems is accurate modeling of the atmospheric turbulence under different conditions. Below are some of the irradiance functions presented in terms of probability distribution functions.

4.1 Lognormal distribution

The lognormal distribution is one of the most widely employed for weak atmospheric turbulence distribution. Here, the irradiance value received at the receiver follows the distribution [9]

$$f_A(a) = \frac{1}{2a\left(2\pi\sigma_x^2\right)^{\frac{1}{2}}} \exp\left(-\frac{(\ln a + 2\sigma_x)^2}{8\sigma_x^2}\right) \tag{1}$$

where μ_x is the mean value of fading and σ_x^2 is the fading covariance.

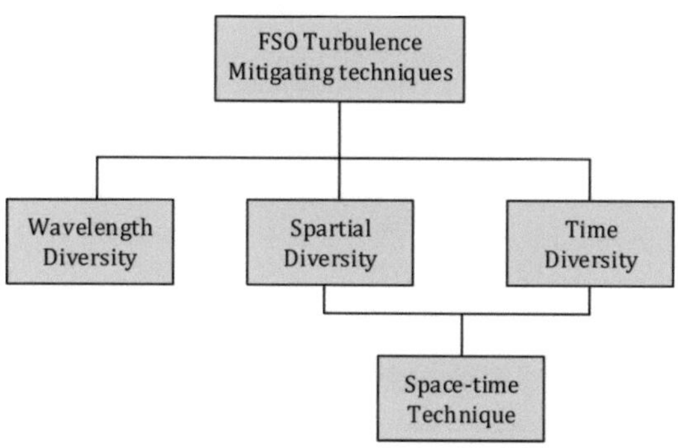

Figure 1.
Some turbulence mitigation techniques in FSO systems.

4.2 *K*-distribution

The K-distribution turbulence model is often used to describe strong atmospheric turbulence conditions (non-Rayleigh sea clutter). For K-distribution atmospheric turbulence model, the probability distribution function $p(I)$ is expressed as [14]

$$p(I) = \frac{2\alpha}{\Gamma(\alpha)}(\alpha I)^{\frac{\alpha-1}{2}} K_{\alpha-1}\left(2\sqrt{\alpha I}\right), I > 0, \alpha > 0 \tag{2}$$

$K_m(\bullet)$ = modified Bessel function of second kind and order m.

4.3 Negative exponential distribution

Negative exponential turbulence model is employed for saturated turbulence cases where the probability distribution function of the received irradiance value is expressed as [15]

$$p(I) = \frac{1}{I_0}\exp\left(\frac{-I}{I_0}\right), I_0 > 0 \tag{3}$$

where I_0 denotes the mean irradiance.

4.4 Gamma-gamma distribution

The gamma-gamma model is very commonly used in FSO communication literatures because it is applicable for a wider range of turbulence conditions. In comparison with measured data, gamma-gamma distribution is effective in describing weak to strong atmospheric turbulence conditions. The PDF is expressed as [16]

$$f\widetilde{H}^{GG}(h) = \frac{2(\alpha\beta)^{\frac{\alpha+\beta}{2}}}{\Gamma(\alpha)\Gamma(\beta)} h^{\frac{\alpha+\beta}{2}-1} K_{\alpha-\beta}\left(2\sqrt{\alpha\beta h}\right) \tag{4}$$

where $K_v(x)$ is the modified Bessel function of the second kind and α and β are the turbulence parameter.

5. Space-time trellis coded coherent FSO

The essence of space-time trellis encoder is to employ mapping functions which are representatives of their trellis diagrams to map binary data to modulation symbols. We design and evaluate the performance of space-time trellis code with two transmit antennas for FSO channel. In order to simplify the design and yet ensuring that there is no jeopardy to the intended MIMO configuration, we represent, for the two transmit antennas, the input bitstream c as [17]

$$c = (c_0, c_1, c_2, \ldots, c_t, \ldots) \tag{5}$$

where c_t, at any instant t denotes a group of two information bits expressed as

$$c_1 = \left(c_t^1, c_t^2\right) \tag{6}$$

As shown in **Figure 2**, the encoder, made up of feedforward shift registers, converts the input bit sequence into a sequence of modulated signals

$x_0, x_1, x_2, \ldots, x_t, \ldots$, where each element x_t of this sequence is the space-time symbol at a given time t.

The output x_t^i of the encoder for the ith transmitter at time t is expressed as [19]

$$x_t^i = \sum_{k=1}^{m} \sum_{j=0}^{v_k} g_{j,i}^k c_{t-j}^k \text{Mod } 2, i = 1, 2 \tag{7}$$

Space-time code (STC) leverages on the features of both time diversity and space diversity to combat turbulence-induced fading in wireless communication systems. RF wireless systems in particular have witness an explosion of interest in the use of space-time coding to improve communication system performance in terms of error control and turbulence mitigation, and FSO communication systems are also witnessing a lot of interest in using this same tool for similar purpose.

In this chapter, we present an adaptive four-state space-time trellis coded coherent FSO system with two transmit lasers, as illustrated in **Figure 3**. Firstly, the error correction performance of the system is complemented by the interleaver, a mechanism put in place to distribute the burst errors—an effect of deep fade, onto different codeword lengths.

Denoting the average SNR as γ, we take the received signal matrix for each codeword C as [20]

$$R = \sqrt{\gamma}\text{CH} + Z \tag{8}$$

where H and Z are the channels and noise matrices, respectively, and H is modeled in terms of the uniformly distributed channel gain phase $\phi_{\mu v}$ and the channel gain amplitude $a_{\mu v}$ as [20]

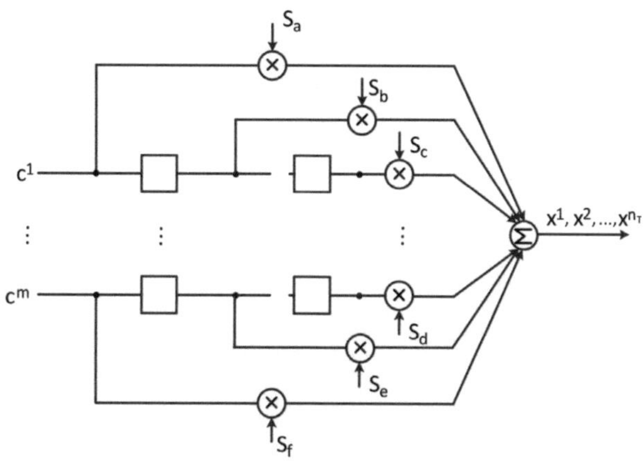

Figure 2.
STTC encoder [18].

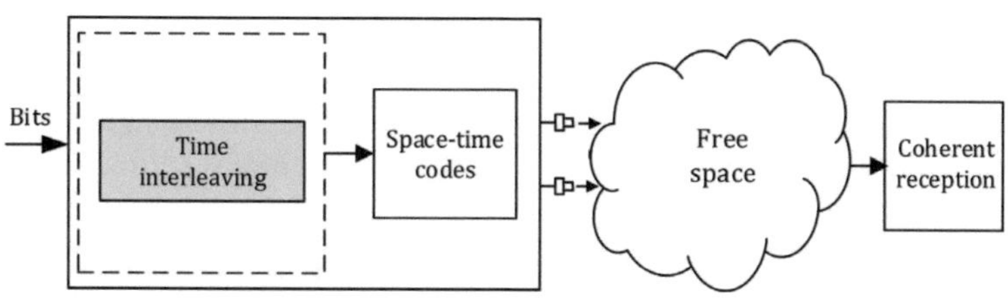

Figure 3.
Space-time trellis coded FSO communication system.

$$[H]_{\mu v} = a_{\mu v} e^{j\phi_{\mu v}} \; 1 \leq \mu \leq 2, 1 \leq v \leq N \tag{9}$$

We begin our analysis using the pairwise error probability (PEP), which is the probability that the decoder erroneously decodes a transmitted STTC codeword **C** as $C' = [c'_0 \ldots c'_{T-1}]$. Then, assuming a gamma-gamma fading distribution as portrayed in Eq. (4), we represent the conditional PEP as [5]

$$P_e(E|H) = Q\left(\sqrt{\frac{\gamma d^2(E)}{2}}\right) \tag{10}$$

where

$$d^2(E) = tr\{H^H E^H E H\} \tag{11}$$

Now, writing a matrix B and its constituent elements as

$$B = \begin{bmatrix} b_{11} & b_{12} \\ b_{12}^* & b_{22} \end{bmatrix} \tag{12}$$

and by equating B with the positive semi-definite matrix $E^H E$ and comparing the elements thereof, where E represents the error matric in the decoding of the codewords and $(\bullet)^H$ denotes the Hermitian transpose function, we write the asymptotic pairwise error probability of the systems as [21]

$$PEP = \frac{\left(\pi a_h^2 \Gamma^2(2\mu) F(\mu, \mu; 1; \xi^2)\right)^N \Gamma\left(2\mu N + \frac{1}{2}\right) \gamma^{-2\mu N}}{2\sqrt{\pi}(b_{11} b_{22})\mu^N \Gamma(2\mu N + 1) \Gamma^{2N}\left(\mu + \frac{1}{2}\right)} \tag{13}$$

where the Gaussian hypergeometric function $F(\bullet)$ is readily computed by using specialized computing functions from libraries of most engineering computing applications or by using fast-converging series [20] as

$$F\left(\frac{t}{2}, \frac{t}{2}; 1; \xi^2\right) = \sum_{n=0}^{\infty} \left[\binom{\frac{t}{2} + n - 1}{n} \xi^n\right]^2 \tag{14}$$

The function Γ in Eq. (13) is a function of the channel parameters α and β; these parameters may be obtained through the Rytov variance, which in turn is a function of the refractive index, the transmission path length between the transmitter and the receiver and the optical wave number [22].

With proper modifications of the values of ξ, Eq. (13) and by extension, Eq. (14), could be modified for general case as well as specific non-orthogonal space-time codes for coherent free space optical communication system. We leverage onto this feature to introduce an adaptive orthogonality controller which adjusts its parameters to any STC supplied thereby not merely eliminating the orthogonality condition as presented in [21] but effectively introduces additional flexibility to the coding scheme.

Readers are to note, however, that several space-time code designs reported for IM/DD FSO communication systems cannot be simply employed for coherent FSO communication systems. This caveat is due to the peculiarities inherent in coherent FSO systems. In addition to this, it should also be noted that in this work, it is assumed that the transmit lasers simultaneously illuminate the receivers with the receivers far away enough from the transmit lasers to assume independent and identically distributed (iid) fading gain.

6. Results and discussion

In this section, the results of the space-time code technique for mitigating turbulence-induced fading in coherent FSO communication systems are presented. Free space optical systems often face the challenge of fading as well as pointing error, and the effect of the latter has been well addressed [23]. The performance of the link under gamma-gamma turbulence is investigated for two transmit lasers, first, with two receivers and then four and six receivers, respectively, as shown in **Figure 4**. Apart from the reduction of the average bit error rate with increase in SNR values, the result shows that at low average SNR, the average performance of the link under the turbulence condition for the different number of receivers are relatively close. However, the difference in performance becomes apparent at higher SNRs as evidenced from SNR 20 to SNR 38.

Although gamma-gamma distribution have been well reported as suitable for modeling weak turbulence as well as strong turbulence scenarios, for the sake of analysis, we employ the values $\alpha = 3.0$ and $\beta = 2.7$. The choice of these values is

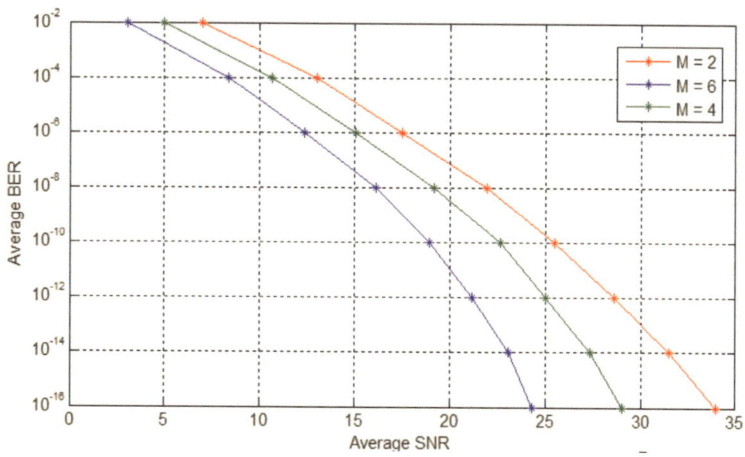

Figure 4.
Performance of coherent FSO link with different receivers.

Coding scheme	Modulation scheme	Detection type	SNR (dB)	References
STBC	OOK	IM/DD	—	[24]
STC variant—no additional constellation extension	PPM	IM/DD	$0 \le \text{SNR} \le 30$	[25]
STBC	OOK	IM/DD with maximum likelihood (ML)	—	[26]
Alamouti-type STC	OOK and PPM	Coherent detection and IM/DD	Additional 3 dB loss relative to BPSK	[27]
STTC	—	IM/DD	$0 \le SNR \le 90$	[28]
Extended Alamouti STC with turbo coding	PPM	IM/DD	$0 \le \text{SNR per bit} \le 30$	[29]
STTC	QPSK	Coherent detection	$0 \le SNR \le 35$	This work

Table 1.
Some coding schemes employed for FSO links.

informed by their popularity in literature as the performance of this work is compared with earlier works in this domain, many of which employed the gamma-gamma distribution parameter above.

FSO communication systems vary in their reception mechanisms as well as modulation techniques. Many works have employed OOK, PPM and QPSK or a combination of these modulation schemes all in a bid to mitigate turbulence-induced fading in FSO links. A few of these works and their corresponding features in comparison to this work are presented in **Table 1**.

Even though the efficiency of space-time codes for turbulence mitigation or error correction in intensity modulated/direct detection FSO systems remains inconclusive in literature, we establish that space-time coding—adaptive space-time trellis codes as in the case of this work, together with inherent potentials of coherent reception for FSO systems—remains a promising solution for free space optical communication systems.

7. Conclusion

For coherent free space optical communication links, we explore the space-time approach to mitigating turbulence-induced fading which thus far remains a serious performance degrading factor for FSO systems. Additionally, as an effort geared towards realizing the promising potentials of coherent free space optical communication systems, we propose an adaptive orthogonality controller for seamless deployment of space-time codes for coherent free space optical communication systems.

Author details

Ojo O. Adedayo[1]*, Oluwafemi B. Ilesanmi[2], Ogunlade M. Adegoke[1] and Ajibade Adedayo[1]

1 Department of Electrical Electronic and Computer Engineering, College of Engineering, Afe Babalola University Ado-Ekiti (ABUAD), Ado-Ekiti, Ekiti State, Nigeria

2 Department of Electrical Electronic Engineering, Faculty of Engineering, Centre for Research in Electrical Communication (CRECO), Ekiti State University, Ado-Ekiti, Ekiti State, Nigeria

*Address all correspondence to: ojoao@abuad.edu.ng

References

[1] Adedayo O, Oluwafemi I, Esan O. Space-time codes in free space optical communications: A case for improved spectral efficiency. The International Journal of Microwave and Optical Technology. 2018;**13**(4):359-368

[2] Kaur G, Singh H, Sappal AS. Free space optical using different modulation techniques—A review. International Journal of Engineering Trends and Technology (IJETT). 2017;**43**(2): 109-115

[3] Mahdieh MHÃ, Pournoury M. Optics & laser technology atmospheric turbulence and numerical evaluation of bit error rate (BER) in free-space communication. Optics and Laser Technology. 2010;**42**(1):55-60

[4] Leitgeb E, Sheikh S, Chlestil C, Gebhart M, Birnbacher U. Reliability of FSO links in next generation optical networks. IEEE ICTCON. 2005:394-401

[5] Garcia-Zambrana A, Castillo-vazquez C, Castillo-vazquez B. Space-time trellis coding with transmit laser selection for FSO links over strong atmospheric turbulence channels. Optics Express. 2010;**18**(6):1293-1300

[6] Tang X, Xu Z. Outage probability of multihop free space optical communications over Nakagami fading channels. 2013. pp. 199–202

[7] Bayaki E, Schober R, Mallik RK. Performance analysis of MIMO free-space optical systems in gamma-gamma fading. IEEE Transactions on Communications. 2009;**57**(11)

[8] Ninos MP, Nistazakis HE, Tombras GS. On the BER performance of FSO links with multiple receivers and spatial jitter over gamm-gamma or exponential turbulence channels. Optik—International Journal for Light and Electron Optics. 2017:1-28

[9] Alimi I, Shahpari A, Sousa A, Ferreira R, Monteiro P, Teixeira A. Challenges and opportunities of optical wireless communication technologies. In: Optical Communication Technology, Pedro Pinho, IntechOpen; 2017

[10] Abou-rjeily C, Member S. Performance analysis of FSO communications with diversity methods: Add more relays or more apertures? IEEE Journal on Selected Areas in Communications. 2015;**33**(9): 1890-1902

[11] Ogunmodede HA, Afullo TJ, Owolawi PA. Performance evaluation of OFDM-FSO communication systems using-QAM and BPSK modulation under log-normal channel. In: The Southern Africa Telecommunication Network and Application Conference; 2016. pp. 98-103

[12] Niu M, Cheng J, Holzman JF. Space-time coded MPSK coherent MIMO FSO systems in gamma-gamma turbulence. In: IEEE Wireless Communications and Networking Conference (WCNC)); 2013. pp. 4266-4271

[13] Kumar P. Comparative analysis of BER performance for direct detection and coherent detection FSO communication systems. In: Fifth Interantional on Communication Systems and Network Technologies; 2015. pp. 369-374

[14] Khalighi MA, Uysal M. Survey on free space optical communication: A communication theory perspective. IEEE Communication Surveys and Tutorials. 2014;**16**(4):2231-2258

[15] Anbarasi K, Hemanth C, Sangeetha RG. A review on channel models in free space optical communication systems. Optics and Laser Technology. 2017;**97**: 161-171

[16] Letzepis N, Guillén A. Outage probability of the Gaussian MIMO free-space optical channel with PPM. IEEE Transactions on Communications. 2009; **57**(12):3682-3690

[17] Rana SM, Alam SMS, Fatema KJ. Code construction and performance evaluation of space time trellis code (STTC) over Rayleigh fading channel. In: Proceedings of the 14th International Conference on Computer and Information Science and Technology; 2011. pp. 22-24

[18] Li Y, Ge Y, Yu J, Dai J, Fu X, Yang E. Improved space-time trellis coded OFDM scheme for frequency selective fading channels. IEEE Canadian Conference on Electrical and Computer Engineering. 2004:1967-1970

[19] Vucetic B, Yuan J. Space-Time Coding. West Sussex, England: John Wiley & Sons Ltd; 2003

[20] Nezampour A, Schober R. Asymptotic analysis of space-time codes in generalized fading channels. IEEE Communications Letters. 2009;**13**(8): 561-563

[21] Bayaki E, Schober R. Performance and design of coherent and differential space-time coded FSO systems. Journal of Lightwave Technology. 2012;**30**(11): 1569-1577

[22] Niu M, Cheng J, Holzman JF. Terrestrial coherent free-space optical communication systems. In: Optical Communication. British Columbia, Canada: The University of British Columbia (UBC); 2012. pp. 169-176

[23] Dubey V, Chandra V. Effect of pointing error on cooperative free space optical communication systems under different channel conditions. In: International Conference on Advanced Networks and Telecommunications System (ANTS); 2015. pp. 2-5

[24] Garcia-Zambrana A. Error rate performance for STBC in free-space optical communications through strong atmospheric turbulence. IEEE Communications Letters. 2007;**11**(5): 390-392

[25] Abou-Rjeily C, Fawaz W. Space-time codes for MIMO ultra-wideband communications and MIMO free-space optical communications with PPM. IEEE Journal on Selected Areas in Communications. 2008;**26**(6):938-947

[26] Wang K, Nirmalathas A, Lim C, Alameh K, Skafidas E. Space-time coded high-speed reconfigurable free-space card-to-card optical interconnects with extended range. The Journal of the Optical Society of America. 2016;**1**(16): 6-8

[27] Abaza M, Mesleh R, Mansour A, Aggoune EM. Spatial diversity for FSO communication systems over atmospheric turbulence channels. In: IEEE WCNC '14 Track 1: PHY and Fundamentals. vol. 1. 2014. pp. 382-387

[28] Mesleh R, Elgala H, Haas H. Optical spatial modulation. Journal of Optical Communications and Networking. 2011; **3**(3):234-244

[29] Khare V, Chadha D. Extended alamouti space time coding scheme with turbo coding for free space optical communication. In: International Conference on Computational Intelligence and Communication Networks; 2011. pp. 366-369

Chapter 7

Spatial Light Modulation as a Flexible Platform for Optical Systems

Cátia Pinho, Isiaka Alimi, Mário Lima, Paulo Monteiro and António Teixeira

Abstract

Spatial light modulation is a technology with a demonstrated wide range of applications, especially in optical systems. Among the various spatial light modulator (SLM) technologies, e.g., liquid crystal (LC), magneto-optic, deformable mirror, multiple quantum well, and acoustic-optic Bragg cells, the ones based on liquid crystal on silicon (LCoS) have been gaining importance and relevance in a plethora of optical contexts, namely, in telecom, metrology, optical storage, and microdisplays. Their implementation in telecom has enabled the development of high-capacity optical components in system functionalities as multiplexing/demultiplexing, switching and optical signal processing. This technology combines the unique light-modulating properties of LC with the high-performance silicon complementary metal oxide semiconductor properties. Different types of modulation, i.e., phase, amplitude or combination of the two, can be achieved. In this book chapter, we address the most relevant applications of phase-only LCoS SLM for optical telecom purposes and the employment of SLM technology in photonic integrated circuits (PICs) (e.g., field-programmable silicon photonic (SiP) circuits and integrated SLM application to create versatile reconfigurable elements). Furthermore, a new SLM-based flexible coupling platform with applications in spatial division multiplexing (SDM) systems (e.g., to efficiently excite different cores in MCF) and characterization/testing of photonic integrated processors will be described.

Keywords: spatial light modulator (SLM), liquid crystal silicon (LCoS) SLM, optical transforms, computer-generated holography (CGH), photonic integrated circuits (PICs), spatial division multiplexing (SDM)

1. Introduction

There has been significant growth in the required capacity of the telecommunication systems, which can be attributed to the proliferation of mobile devices, bandwidth-intensive applications, and services [1–3]. As a result, a significant increase in the broadband connections as well as the related multimedia traffic on a yearly basis [4–6] has been progressing. Moreover, the traffic explosion has been one of the challenges being faced in telecommunication systems [2, 7]. Also, it has been observed that the traditional electronic media which are based on copper are unable to meet the system requirements majorly in terms of bandwidth and latency [5, 8–10].

To address the challenges, optical fiber-based transport systems have been employed in different fields of communication systems as viable and reliable solutions. The widely employed optical transport systems are based on single-mode fiber (SMF). To enhance the capacity of single-core SMF, advanced modulation formats and wavelength division multiplexing (WDM) are normally employed. However, the growing demand for further video/image storage capacity and the increase in cloud service adoption, which is as a result of numerous smartphones and other Internet-based gadgets, have led to research on solutions for effective bandwidth optimization [11, 12]. This is due to the fact that the conventional SMF-based transport systems have been observed to be approaching Shannon's limit [13] and the achievable maximum capacity will not be sufficient to support the envisaged massive connection demanded by the next-generation networks [5, 14–16]. Besides the capacity that is expected to be saturated around 100 Tbit/s owing to the physical limits, the conventional SMF schemes with WDM might be unable to meet the power consumption, spatial efficiency, and cost requirements of the communication systems [16, 17].

There has been considerable attention on multicore fiber (MCF) as a feasible solution capable of addressing the capacity limit of a conventional SMF-based scheme [15, 18, 19]. For effective implementation of MCF, the research community has been working diligently on improved cost-efficient and scalable networking infrastructure solutions. A notable optical transport scheme that can exploit the space dimension in order to address the optical system capacity crunch and improve the system performance is spatial division multiplexing (SDM) [19]. Moreover, it has been observed that MCF is an efficient and main enabling technology for the SDM systems [16, 18]. Apart from the MCF, SDM implementation for multimode fiber (MMF) has also been attracting significant attention [17, 18]. Nevertheless, the MCF implementation is susceptible to and can be constrained by the transmission impairments such as nonlinearities and inter-core cross talk (XT) between signals at the neighboring cores that may be presented via multiple optical paths. This may have a significant effect on the system performance regarding the transmission range and the network size [16, 17]. Furthermore, the extent of the presented performance degradation by the transmission impairments varies with the MCF fiber types (i.e., 3-core, 7-core, 13-core, 19-core, 37-core, and 61-core) [16, 20]. A practical solution for addressing the MCF implementation challenges is spatial light modulation.

Spatial light modulators (SLM) can be employed for exciting different cores and/or modes in order to mitigate the transmission impairments introduced by multiple optical paths, as it enables arbitrary removal or addition of channels with the aid of software, i.e., implementation of a diffractive optical element by computer-generated holograms (CGH). Due to the SLM support for dynamic reconfiguration of optical wave fronts, it can be employed for core and mode multiplexing and demultiplexing [5, 21, 22]. In addition, the use of silicon photonic (SiP) onboard transceivers that are coupled on the MCF for supporting transmissions has been shown to be promising. This is due to the fact that there is no need for fan-in/fan-out or core pitch conversion devices that may give rise to further system complexity [16].

Optical communication evolution has brought about the emergence of improved photonic integrated circuits (PICs) that present economic and sustainable alternative to data transmission [9]. Therefore, it is expected to be an enabling technology, capable of contributing significantly in a number of fields [8]. As a result, various benefits are offered, such as small weight and volume, low power consumption, high mechanical and thermal stability, and the ease of assembling a number of complex systems.

PIC can be generally characterized as a multiport device with an integrated system of optical elements such as attenuators, modulators, multiplexers, detectors, lasers, and optical amplifiers that are embedded in a single chip using a waveguide (WG) architecture [23]. However, it has been observed that optical component testing is very challenging and time-consuming as well [24]. This can be attributed to the required tight three-dimensional (3D) alignment tolerances, to ensure accurate light coupling. Hence, with notable development and growing adoption of PIC in the communication networks, advanced methods are imperative for an accurate PIC performance testing as well as characterization. As aforementioned, based on the support for dynamic reconfiguration of light, SLM can be employed for optical PIC testing and characterization, by exploring this feature in the feeding and/or receiving the optical signal into the PIC [8, 23].

In this chapter, we focus on the most pertinent applications of phase-only liquid crystal on silicon (LCoS) SLM for optical telecom purposes and the employment of SLM technology in PIC, e.g., field-programmable silicon photonic circuits and integrated SLM application to create versatile reconfigurable elements. Furthermore, a new SLM-based flexible coupling platform for applications in SDM systems and characterization/testing of photonic processors will be presented.

In Section 2, SLM working principle and their applications in telecom are addressed in more detail. The applied methodology (i.e., algorithms and experimental setup) to create a diffraction optical element through the implementation and optimization of CGH is described in Section 3. In Section 4, discussion on the obtained results from the experimental implementation of CGH for SDM and PIC applications is presented. An overall conclusion regarding the employment of the SLM technology as a flexible platform for optical systems is provided in Section 5.

2. Spatial light modulator (SLM)

Optical signal processing has been providing relevant solutions to convert data into spatially modulated coherent optical signals with SLM devices, allowing the effective implementation of digital holograms [25]. One of the most useful properties of the hologram is its ability to control phase and amplitude of light in the far field. The Fourier transform describes the relationship between a hologram (near field) and its corresponding replay field (far field). The far field can be formed at the focal point of a positive lens or an infinite distance from the near field plane in free space [25, 26], as depicted in **Figure 1**.

Holograms can reproduce waveforms from an existing object. With digital advances and optical signal processing, it is possible to numerically calculate interference patterns to generate completely synthetic wave fronts of arbitrary form. These interference patterns can have different denominations, such as CGH, diffractive optical elements (DOE), phase/amplitude masks, diffractive grating, etc. [26]. All operate in the principle of diffraction, so it is somehow an arbitrary choice of terminology.

The SLM is a device that can be used to modulate light in accordance with a fixed spatial (pixel) pattern and can be programed electrically. Usually, it can be exploited for incident light phase and/or amplitude control. Subsequently, phase-only, amplitude-only, or the combination of phase-amplitude can be readily realized with SLM. There are a number of modulation mechanisms that can be employed. One of the attractive and widely used ones is electro-optical SLM. The

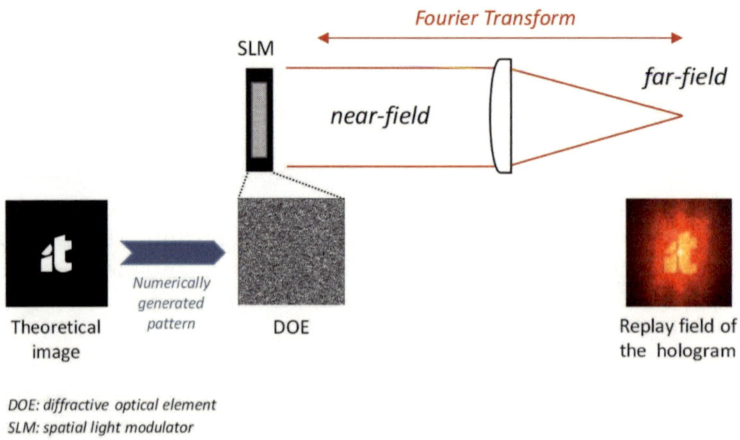

Figure 1.
Diagram of a Fourier transform through a positive lens. A complex design pattern (diffractive optical element (DOE)) is provided to the SLM to generate the expected hologram in the replay field (far field).

modulation material of electro-optical SLM is liquid crystal. Similarly, a liquid crystal SLM has a microdisplay being employed for the incident light modulation and collection. This can be realized in a transmissive form using a liquid crystal display (LCD) SLM technology or in a reflective form with LCoS SLM technology. One of the leading features of the modulators is in the liquid crystal molecule alignment. Typically, this can be either vertical, parallel, or with twisted formation. Consequently, with suitable polarizing optics, this influences the incident light beam properties that can be effectively changed, i.e., amplitude, phase, or their combination [8, 23, 27].

It should be noted that the common hologram generation techniques cannot arbitrarily modulate the beam phase and amplitude concurrently [26, 27]. Therefore, it is unrealistic to basically address the desired pattern inverse Fourier transform into the far field and replicate the resulting amplitude and phase distribution directly on the SLM. As a consequence, the employment of optimization algorithms is highly recommended for evaluating the best potential hologram within the device constraints, e.g., the best pixel distribution in which each pixel will be able to take only one of two states that correspond to a 0 or π phase shift [23, 26].

The nematic LCoS technology is a type of SLM with phase-only modulation capability. Moreover, it is an electrically addressed reflection modulator category in which a direct and accurate voltage controls the liquid crystal and the light beam wave front can be modulated as well [28, 29]. An example of an LCoS SLM is illustrated in **Figure 2** [23]. The LCoS SLM can be employed as a diffractive device for reconstructing images from CGH [30].

CGH can be employed for different communication purposes and has been gaining application in indoor visible light communication systems [31]. Furthermore, suitable holograms can be readily generated by employing a variety of optimization techniques such as iterative Fourier transform algorithm (IFTA) [5, 32–34]; linear Fourier transform (i.e., linear phase mask) [5, 18, 23, 35]; simulated annealing [36]; and Gerchberg-Saxton algorithm [37]. The employment of the SLM as a diffractive device for reconstructing images from CGH permits the light beam wave front to be modulated [8, 23].

As aforementioned, LCoS displays have been gaining significant recognition as promising microdisplays for various types of SLM applications. Similarly, they possess attractive and significant features like very high spatial resolution and

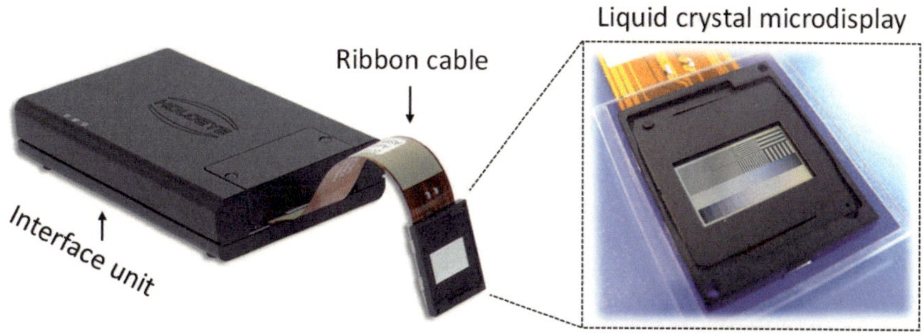

Figure 2.
LCoS SLM Pluto phase modulator from Holoeye© 2018 Holoeye photonics AG.

light efficiency [38]. Due to this, they have been applicable in a plethora of optical contexts such as communication, reconfigurable interconnects [39], storage [40], diffractive optics [41], metrology [42], and quantum computing [43]. They are also applicable in the wave shaper technology for optical signal processing and monitoring [44]. In addition, other advantages of the LCoS are that it is highly cost-effective and can as well be flexibly programmed. This helps in supporting a number of additional functions like group delay ripple compensation, wavelength filtering, and chromatic dispersion compensation. Besides, it can aid in ensuring variable attenuation for individual wavelength channels as well as output ports. Consequently, LCoS device offers a cost-effective and promising solution for the wavelength selective switch (WSS) [40].

The LCoS microdisplay SLM has a good implementation history in the WSS systems. Its employment in the WSS system core component can be attributed to a number of advantages such as larger spatial bandwidth, more port availability, and enhanced resolution, as well as the system miniaturization. The WSS systems have been exploited in the reconfigurable optical add/drop multiplexers (ROADM) in WDM optical networks. It has been observed that ROADM is one of promising schemes that can be employed to improve on the traffic capacity of the existing and future telecommunication systems [40, 45]. Moreover, in communication networks, the ease of adding or dropping the wavelength is essential. They can ensure effective information access or rerouting to another appropriate path in the network. It should be noted that WSS is the ROADM sub-system that has been extensively employed in optical switch applications. In addition, microelectromechanical systems (MEMS) [46] and WSS based on LCoS [47] have been commercialized for different applications. Also, WSS by means of LCoS operates on the principle of "disperse and select," in which the inward bound WDM channels are dispersed into a distinct wavelength channel and subsequently relayed by LCoS through programmable grating patterns. This is in an attempt to facilitate an add and drop function. It is envisaged that the next-generation ROADM will hold different attractive features such as directionless, colorless, and contentionless in order to improve the system performance [40].

Furthermore, LCoS technology can also be employed in flex grid that has been considered as the major feature for the next-generation networks [40, 48]. It should be noted that the traditional fixed grid with 50 GHz spacing standardized by the International Telecommunication Union (ITU) Telecommunication Standardization Sector (ITU-T) possesses a number of challenges. The fixed grid has been observed to bring about the optical spectra being inefficiently used. Besides, it constrains the system transmission capacity considerably. On the other hand, the flex grid implementation enables the use of different modulation formats and their coexistence on a shared infrastructure. They can also be densely and efficiently multiplexed which aids the optical networks, not only to extend the reach but also the per channel bit rate. It has also been envisaged that implementation of WSS and SDM will significantly help further in extending the network reach and capacity [40].

3. Methodology

The adopted methodology to implement the SLM flexible platform for optical systems can be subdivided into two main sections: (i) the algorithms employed for the CGH generation and optimization methodology (in Section 3.1) and (ii) the SLM framework setup implementation with application in SDM systems and characterization/testing of PIC (in Section 3.2).

The framework ability to improve the overall alignment process and excite different cores of a MCF, can provide a valuable contribution for the impairment mitigation of the system optical path, which can relax digital signal processing (DSP) equalization requirements of the SDM system [5, 18, 22, 34, 49].

Furthermore, its use as a flexible platform for feeding photonic integrated processors was also explored for the characterization/test of PICs, and results have been presented for its implementation as a parallel implementation of the Haar transform (HT) image compression algorithm [8, 18, 23].

3.1 CGH pattern establishment

Holography is a 3D-based display system that comprises exploiting diffraction and interference for recording and reconstructing optical wave fronts. Moreover, computer-generated holography is an effective technique that is appropriate for a broad variety of displays such as two-dimensional (2D), volumetric, autostereoscopic, stereoscopic, and true 3D imaging. It is remarkable that the CGH is becoming feasible due to the emergence of progressively powerful computers that prevents the conventional interferometric recording step in the formation of hologram [50]. In addition, the CGH can be viewed as a phase mask with spatially variable transmittance or a diffractive optical element that can be readily displayed on the devices such as SLM, which are capable of diffracting light [5, 51]. Also, the information that needs to be transformed is presented to an optical system, through the SLM. This is effected with a suitable phase mask for the concerned input function [25].

From a set of different available techniques for the generation of CGH (e.g., IFTA, linear Fourier transform, simulated annealing, and Gerchberg algorithm, as described in Section 2), in our SLM framework, higher focus was given to the linear Fourier transform principle for the calculus of the numerical interference patterns to generate the holograms (CGH). This decision was mainly due to the intensive computational requests and high power loss (up to 9 dB [26]) associated with the implementation of the simulated annealing and Gerchberg-Saxton algorithms and additional computational cost of IFTA when compared to linear phase mask.

Thus, a simplified approach based on the implementation of a linear phase mask generation (in Section 3.1.1) and the development of a new iterative algorithm experimentally driven for CGH effective optimization (in Section 3.1.2) is proposed and tested.

All algorithms were developed and implemented in MATLAB© [52].

3.1.1 Linear phase mask CGH

A linear phase mask can be described as a numerical information transformation (in the Fourier domain) of the input function of interest [25], which can be introduced into the optical system through an SLM.

The CGH implemented with a linear phase mask can be expressed in the frequency domain as expressed in Eq. (1) [5, 23], where f_x and f_y denote the spatial frequency vector components that correspond to the image to be generated in both X and Y axes, respectively, and c_x and c_y represent the horizontal and vertical tilt parameters, respectively.

$$M_{linear} = M(f_x, f_y) = -2\pi(c_x f_x + c_y f_y) \tag{1}$$

A collimated Gaussian beam with transverse profile S_{in} is imaged onto the SLM via a lens, Eq. (2), where (x_0, y_0) offer the horizontal and vertical position, respectively, and (w_x, w_y) represent the width and the height of the beam, respectively, as depicted in **Figure 3** [8].

$$S_{in} = exp\left(-\left(2\frac{x - x_0}{w_x\, log(\sqrt{2})}\right)^2 - \left(2\frac{y - y_0}{w_y\, log(\sqrt{2})}\right)^2\right) \tag{2}$$

With the adoption of Fraunhofer approximation, the Fourier transform is produced on the SLM plane, $fft(S_{in})$. Afterward, the subsequent illumination profile is multiplied with the phase mask, $e^{iH_{mask}}$. The resultant signal is then Fourier transformed via the second lens by means of an inverse Fourier transform to achieve the output signal S_{out}, which can be defined as Eq. (3) [5, 8]:

$$S_{out} = ifft(H(fft(S_{in}))) \tag{3}$$

A graphical user interface (GUI) was also developed to test different masks to be applied to the SLM device [18] (see **Figure 4**).

Different phase masks can be attained by adjusting the different available parameters from the developed GUI. For the *Input Beam* GUI panel the following

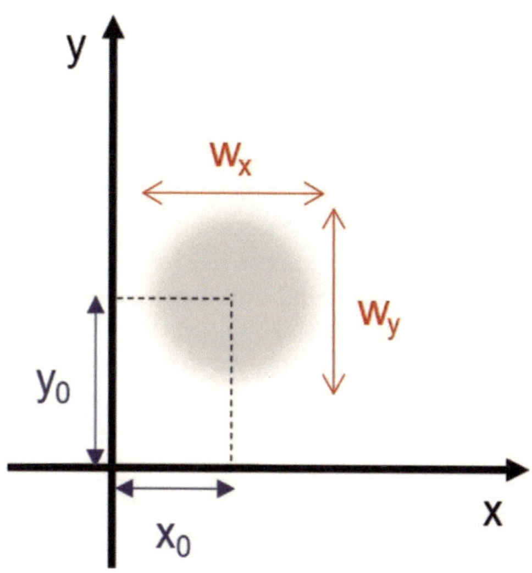

Figure 3.
Cartesian coordinate system description of the parameters (x_o, y_o) and (w_x, w_y) employed for the input beam S_{in} estimation [8].

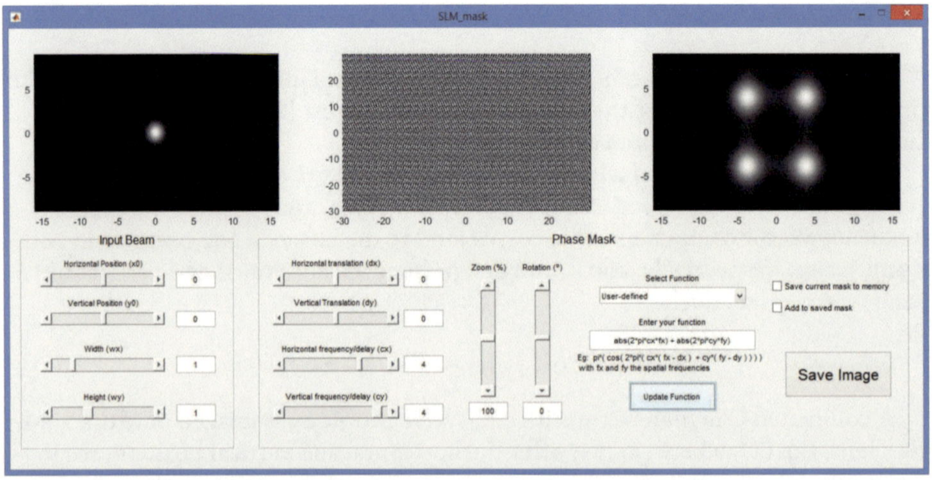

Figure 4.
GUI SLM_mask to generate the different phase masks applied to the SLM [18].

input parameters are available: (i) horizontal position (x_0); (ii) vertical position (y_0); (iii) width of the beam (w_x); and (iv) height of the beam (w_y) (see GUI panel *Input Beam* in the **Figure 4**).

$$S_{out} = ifft(H(fft(S_{in}))) \tag{4}$$

The Phase Mask GUI panel offers the correspondent input parameters: (i) horizontal translation (d_x); (ii) vertical translation (d_y); (iii) horizontal frequency delay (c_x); (iv) vertical frequency delay (c_y); (v) percentage of zoom (%); (vi) rotation in degrees (°); and (vii) selection of three possible input functions, i.e., sinusoidal Eq. (4), linear Eq. (1), or defined by the user (user-defined). The option to save or replace the phase mask file is also made available, as depicted in the Phase Mask GUI panel from **Figure 4**.

The implemented scripts and GUI were written in MATLAB© [52].

3.1.2 Optimization of the linear phase mask CGH

In an effort to realize the hologram that can suitably replicate the output signal, we estimated the hologram of the beam through the image phase-only information of the generated hologram. Thus, a first linear phase mask is generated to produce the expected initial field, i.e., the input function of interest.

Since a phase-only SLM does not permit the inverse Fourier of the desired pattern to be addressed into the far field and replicated into the resultant distribution of amplitude and phase on the SLM directly. It is quite demanding to generate a CGH with guarantees for the light to be spatially modulated with the required accuracy and resolution. To address these challenges and obtain the desired hologram with an error factor $\delta \leq 10\%$, we implemented an iterative algorithm to optimize the generation of the linear phase mask. Also, the error factor threshold was set so as to prevent an infinite loop in the adopted optimization algorithm, while guaranteeing that the output result has an accuracy $\geq 90\%$.

The algorithm was implemented to generate a hologram that replicates the output of the four waveguides (WG) of an optical chip for data compression proposes [8, 23, 53]. A hologram of four beams was calculated by a phase-only superimposition of four independent holograms generated by Eq. (1). Then, the corresponding linear transformations in the Fourier domain provided in Eq. (5) and Eq. (6) were applied [8, 23]:

$$H = \angle\left(e^{iH_1} + e^{iH_2} + e^{iH_3} + e^{iH_4}\right) \tag{5}$$

$$H_1 = exp\left(i2\pi\left(c_{x1}f_x + c_{y1}f_y\right)\right) \tag{6}$$

The block diagram of the employed algorithm is given in **Figure 5**, and the major steps of the algorithm are enumerated as follows [23]:

i. Generate a first linear phase mask to produce the expected initial field based on Eq. (5).

ii. Initially set the four values $a_{1\text{-}4}$ to 1, from $H = \angle\left(a_1 e^{iH_1} + a_2 e^{iH_2} + a_3 e^{iH_3} + a_4 e^{iH_4}\right)$.

iii. Acquire the replay field from the hologram generated by SLM ($I\ SLM$) with a camera, and feed this data to the algorithm.

iv. Calculate the difference between the hologram generated and the initial field expected, defined as error factor: $\delta = |I_{SLM} - I_1| \le 0.1$.

v. If the condition $\delta \le 0.1$ is not satisfied, repeat steps ii–iv by iteratively adjusting the values of $a_{1\text{-}4}$ to compensate for the error factor.

The developed algorithm in MATLAB® is capable of controlling both SLM and camera hardware, providing a dynamic experimentally driven algorithm for effective CGH optimization.

The error factor (δ) is defined to quantify the generated hologram deviation from the optical chip anticipated output [8, 23].

3.2 CGH generation setup

The employed SLM is a reflective LCoS phase-only type, and its model is PLUTO-TELCO-012. It can operate within the wavelength range of

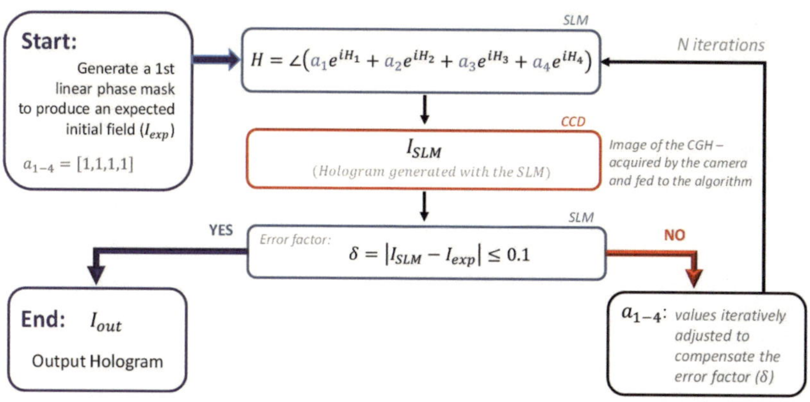

Figure 5.
Block diagram of the algorithm applied for the optimization of the linear phase mask CGH [23].

1400–1700 nm. Additionally, an active area of 15.36 × 8.64 mm^2, a pixel pitch of 8.0 μm, a 92% fill factor, and 80% reflectivity are employed for displaying the generated hologram.

Two different setup arrangements were implemented to create CGH for SDM (e.g., MCF) and PIC applications.

3.2.1 CGH setup for SDM

Setup alignments were carried out, using a red laser of 637 nm (power 70 mW, SM fiber-pigtailed laser diode), a collimator, two lenses, a charge-coupled device (CCD) image sensor, and the LCoS-SLM. After the alignments, an MCF of 10 m of length and a bit error rate (BER) tester were introduced in the setup, as depicted in **Figure 6**.

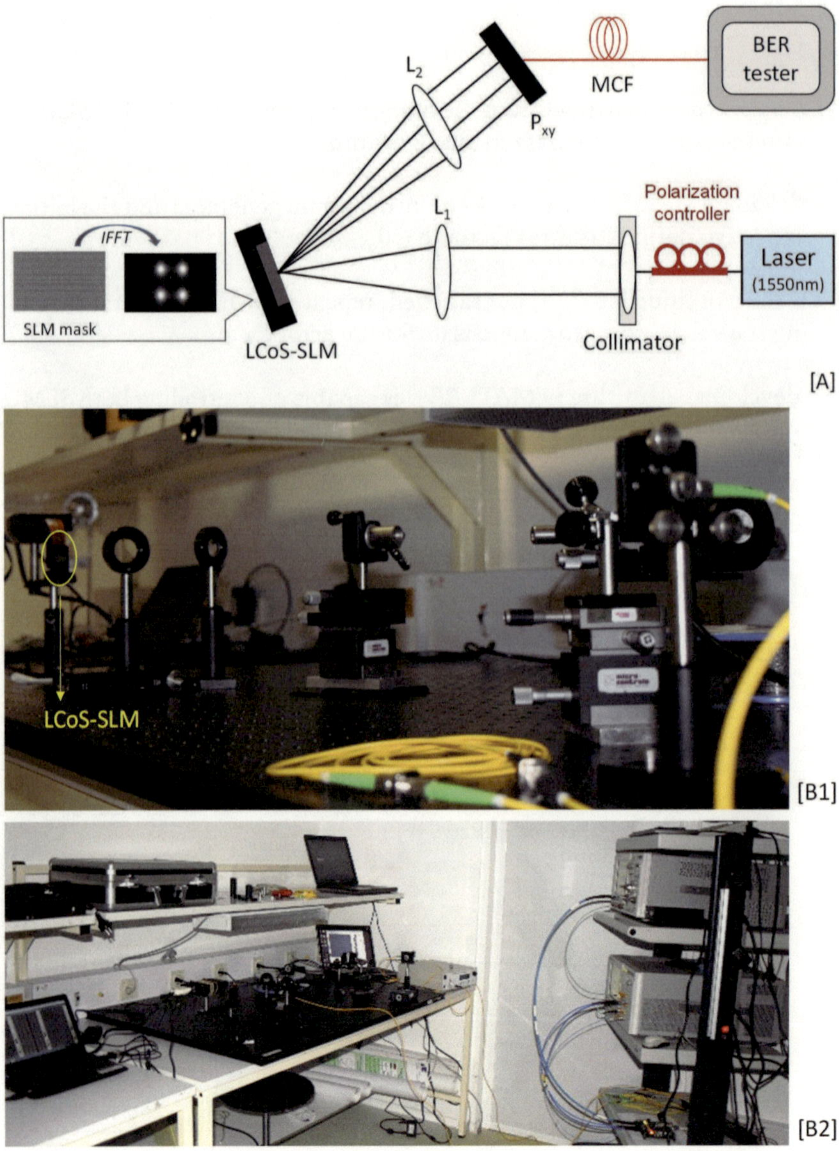

Figure 6.
[A] Setup diagram of the SLM platform for MCF applications. [B1, B2] photographs of the setup.

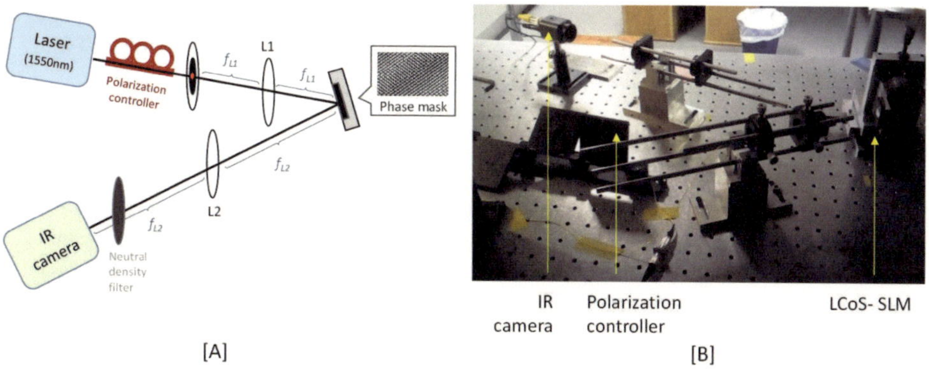

Figure 7.
[A] Hologram reconstruction scheme using an infrared (IR) laser of 1550 nm, a polarization controller, lens L1, an LCoS-SLM, lens L2, and an IR camera. [B] Photography of the presented setup.

The MCF contained four cores arranged in a quadrangular lattice pattern, with a side length of 36.25 μm and attenuation @1550 nm of 0.45 dB/km.

The nonreturn-to-zero (NRZ) signal was generated by a pattern generator (Agilent N4901B) using a pseudorandom binary sequence (PRBS) 2^{31}–1. This signal was injected to the tunable direct modulator laser to create 10 Gb/s optical signal. After the MCF, the signal was detected by avalanche photodiode (APD) receiver inside the small form-factor pluggable (SFP) transceiver.

3.2.2 CGH setup for PIC

In an effort to eliminate the phase distortion and enable the full Fourier transform scale by the focal length (f) factor, the optical system is designed based on the $4f$ system configuration. The implementation is the basis of a low distortion optical system.

The setup consists of devices such as two lenses (AC254–050-C-ML, AR coating 1050–1620 nm) L1 and L2 with a focal length of 75 and 250 mm, respectively; polarization controller; an infrared (IR) laser of 1550 nm (wavelength); a near-infrared (1460–1600 nm) camera (sensing area, 6.4 × 4.8 mm; resolution, 752 × 582; pixel size, 8.6 × 8.3 μm) for capturing the produced hologram; and a neutral density filter, to prevent saturation in the camera acquisition (see **Figure 7**) [8, 23].

4. Results and discussion

In this section, we present the experimental CGH results obtained for the SDM and PIC applications.

4.1 SLM platform for MCF

A BER of 1.2×10^{-3} was measured in the experiment described in Section 3.2.1. Test result shows an error-free transmission below the BER limit of 3.8×10^{-3} (7% hard-decision FEC) [54, 55] threshold.

Thus, the SLM framework was able to properly function as a spatial coupling interface between the SLM generated pattern and the MCF cores. The platform allows an easy adjustment of the generated phase mask (CGH), contributing to an effective dynamic optimization of the MCF fiber transmission [18].

Future work will be performed in order to optimize the current convergence, namely, improve the optical system components (e.g., lenses and collimator) and the implemented phase masks [18].

4.2 CGH for PIC applications

An integrated approach for compression applications implemented in an indium phosphide (InP) optical chip was fabricated to realize a Haar wavelet transform [53]. The HT is a wavelet-based method with promising attributes for compression transformation techniques. Their application in image processing and pattern recognition due to its simple design, fast computation power, and efficiency can be easily realized by optical planar interferometry [53, 56, 57].

The HT operations include low-pass (L) and high-pass (H) filters applied over one dimension at a time. This filtering operation corresponds to the calculation of the average between two neighbors' pixel values (LP) or the difference between them (HP) [57]. The HT is implemented with a two-level network composed by three asymmetric adiabatic couplers (AAC) 2 × 2, reproducing the required operations, i.e., the average (sum) and the difference (subtraction) between the optical input pair [53]. The 2D HT can be decomposed in four sub-bands, LL, LH, HL, and HH [57]. The LL gives the data compressed.

In the optical chip (or PIC), these four sub-bands can be extrapolated from the four output WG at the end of the three AAC network, as depicted in **Figure 8**.

The optical chip is composed of four distributed feedback (DFB) source lasers (L_1–L_4), three asymmetric adiabatic coupler (AAC_1–AAC_3), six positive-intrinsic-negative (PIN) photodiodes for network monitoring, six MMI splitters 1 × 2, one MMI splitter 2 × 2, and two spot size converters (SSC), at the correspondent HT network output LL (compression) and HH. Further details about the PIC can be found elsewhere [23, 53].

In the described SLM framework, the hologram is generated in an attempt to create the beam profile in the first order of diffraction when being displayed on the SLM. The CGH is expected to reproduce the four WG outputs of the PIC implementing the HT [23, 53] (see **Figure 8**).

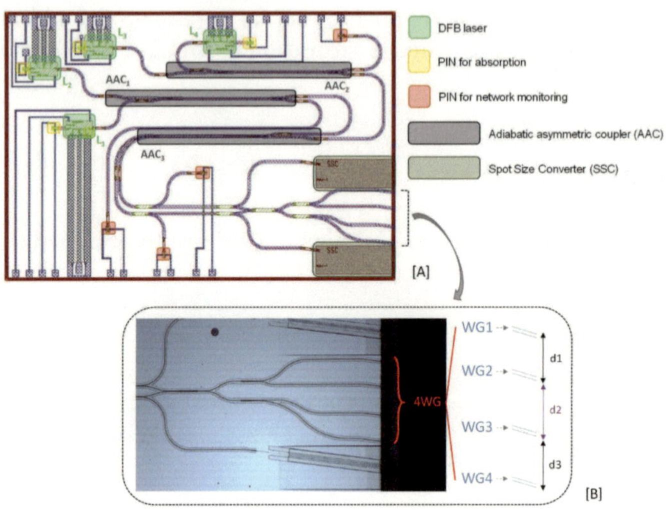

Figure 8.
[A] Design architecture of the PIC for data compression based on HT. [B] Measurements of the distance between the four WG at the end of the two-level compression network of the PIC [23].

In **Figure 9**, we present the obtained image from the hologram replay field generated with the initial (I_1) and optimized (I_{out}) CGH [8, 23].

The analysis of the obtained replay field images can be described by the steps described below:

i. Calculate the intensity integration of the image matrix, i.e., the sum of all elements along each line of the image matrix, depicted as S_{raw}.

ii. Application of the Savitzky–Golay (SG) filter to smoothen the intensity integration signal obtained in step (1), depicted as S_{SG}.

iii. Implementation of a first-order Gaussian fit curve to the filtered signal, depicted as *Gauss fit*.

iv. Extraction of Gaussian parameters to calculate the distances between the four beams (obtained from the CGH) and compare with the expected results (d_1, d_2, and d_3 from the optical chip).

Results from the integrated intensity profile of the replay field after the CGH optimization, i.e., after step (3), are presented in **Figure 10**.

The distance between the four beams was calculated from the center position of each beam profile, given by the Gaussian fit coefficient corresponding to the position of the center of the peak. The coefficients were obtained with 95% confidence bounds [23]. The deviation values (δ) between the generated holograms (i.e., initial I_1 and optimized I_{out}), when compared with the expected output of the optical chip (i.e., d_1, d_2 and d_3 from **Figure 8**), are presented in **Table 1** [23].

The measured power of the beams obtained by the integration intensity profiles is depicted in **Table 2** [23].

An improved hologram is achieved with the optimization of the linear phase mask CGH, i.e., with a reduction of up to 11% in the error factor (between

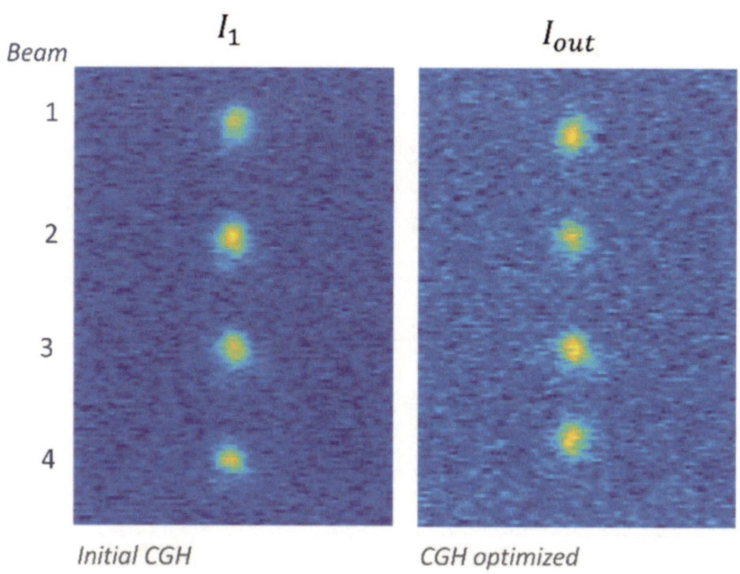

Figure 9.
Hologram replay field obtained by the IR camera with an (i) initial hologram (left figure) and (ii) optimized hologram (right figure).

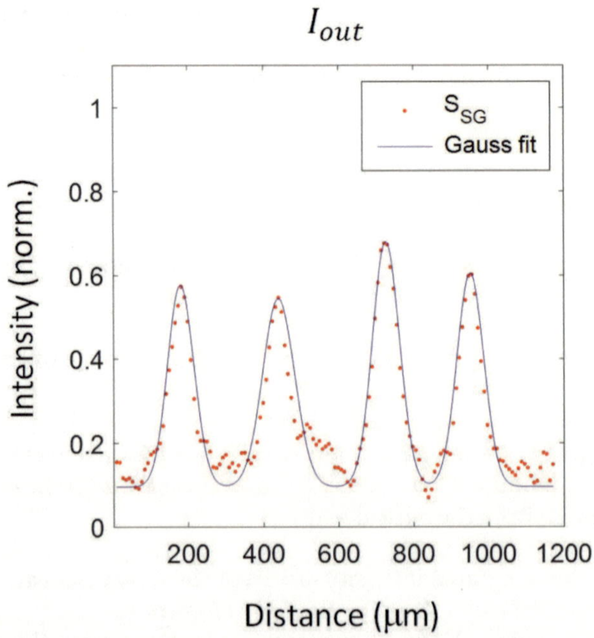

Figure 10.
Gaussian fit (Gauss fit, blue line) of smoothed integrated intensity signal from the replay field image (S_{SG}, red dots) of final optimized CGH [23].

	Initial CGH (%)	Optimized CGH (%)
δ_{d1}	19.76	7.48
δ_{d2}	1.96	2.90
δ_{d3}	14.31	9.44

Table 1.
Error factor (δ) values for d1, d2, and d3.

Beam	Initial CGH (u.a.)	Optimized CGH (u.a.)
1	6.30	5.12
2	8.21	5.78
3	7.18	6.37
4	7.69	5.51
Mean	7.35 ± 0.81	5.69 ± 0.52
Std (%)	11.17	9.14

Table 2.
Integration of the intensity profiles for the four beams.

initial and optimized holograms). Nonetheless, the loss of 1.1 dB identified on the mean beam power for the optimized CGH, an improved equalization between the beams was observed, with a 2% reduction in the standard deviation [23].

Algorithm improvements should be implemented to mitigate the power discrepancies between the four beams and optical artifacts associated with the diffraction of light, with the objective of mitigating the signal loss at the output of the optical chip.

A possible approach to correct some of this artifacts can be the application of the Gerchberg-Saxton [37] or simulated annealing [36] algorithms; nonetheless, due to the power loss (up to 9 dB [26]) associated with these approaches, they were not addressed in this implementation [23].

The phase mask that replicates the expected output of the PIC optical operation can be used to multiplex/demultiplex the obtained result. Furthermore, a phase mask, which addresses the HT operations, can also be applied to invert the compression induced by the HT (optically implemented in the PIC all-optical network with three AAC). Thus, a proof of concept of the PIC operation through the SLM coupling framework is expected [8, 23].

5. Conclusion

LCoS SLM technology implementation has been gaining importance in optical system applications, like telecom with the development of high-capacity optical components in system functionalities as switching (in ROADM), multiplexing and demultiplexing, and optical signal processing. In this chapter, a proof of concept on the implementation of a new SLM-based flexible coupling platform has been provided. We have also explored its implementation for applications in SDM systems and PIC characterization/testing. Furthermore, optimized methodologies to generate the CGH were developed and implemented. Main results include (i) BER = 1.2×10^{-3} for a SDM system, i.e., the use of the SLM to efficiently excite the different cores of a MCF, and (ii) CGH ($\delta \leq 1.5\%$) to feed/receive the output of an optical chip for data compression based on the HT. The demonstrations pave the way for the potential use of the SLM flexible platform in the development of multidimensional optical systems, by providing a versatile optical method which can overcome impairments introduced by the optical path in a MCF (e.g., by improving the setup alignment and excitation of different cores in MCF) and deliver a more robust optical methodology to assess and test photonic processors (e.g., offering a proof of concept of the PIC HT operation).

Acknowledgements

This work is funded by Fundação para a Ciência e a Tecnologia (FCT) through national funds under the scholarship PD/BD/105858/2014. It is also supported by FCT/MEC/MCTES and when applicable co-funded by FEDER – PT2020 partnership agreement under the project COMPRESS – PTDC/EEI-TEL/7163/2014 and the project UID/EEA/50008/2019. The authors acknowledge PICadvanced for its collaboration.

Author details

Cátia Pinho[1,2*], Isiaka Alimi[1,2], Mário Lima[1,2], Paulo Monteiro[1,2] and António Teixeira[1,2]

1 Instituto de Telecomunicações (IT), University of Aveiro, Aveiro, Portugal

2 Department of Electronics, Telecommunications and Informatics (DETI), University of Aveiro, Aveiro, Portugal

*Address all correspondence to: catiap@ua.pt

References

[1] He J et al. A survey on recent advances in optical communications. Computers and Electrical Engineering. 2014;**40**(1):216-240

[2] Ericsson. Ericsson Mobility Report. 2018

[3] Winzer PJ. Spatial multiplexing in fiber optics: The scaling of metro/core capacities. Bell Labs Technical Journal. 2014;**19**:22-30

[4] Ayyash M et al. Coexistence of WiFi and LiFi toward 5G: Concepts, opportunities, and challenges. IEEE Communications Magazine. 2016;**54**(2):64-71

[5] Pinho C, Shahpari A, Alimi I, Lima M, Teixeira A. Optical transforms and CGH for SDM systems. In: 18th International Conference on Transparent Optical Networks (ICTON 2016). 2016. pp. 1-4

[6] Alimi IA, Monteiro PP, Teixeira AL. Outage probability of multiuser mixed RF/FSO relay schemes for heterogeneous cloud radio access networks (H-CRANs). Wireless Personal Communications. 2017;**95**(1):27-41

[7] Alimi IA, Monteiro PP, Teixeira AL. Analysis of multiuser mixed RF/FSO relay networks for performance improvements in cloud computing-based radio access networks (CC-RANs). Optics Communication. 2017;**402**:653-661

[8] Pinho C et al. Flexible platform for feeding photonic integrated processors. In: The Thirteenth Advanced International Conference on Telecommunications (AICT 2017). 2017. pp. 1-4

[9] Grand View Research. Photonic Integrated Circuit (IC) Market Size Report. 2016

[10] Credence Research. Photonic Integrated Circuits Market. 2016

[11] Desurvire E et al. Science and technology challenges in XXIst century optical communications. Comptes Rendus Physique. 2011;**12**(4):387-416

[12] Ip E, Kahn JM. Increasing optical fiber transmission capacity beyond next-generation systems. In: LEOS 2008. 21st Annual Meeting of the IEEE Lasers and Electro-Optics Society 2008. pp. 606-607

[13] Shannon CE. A mathematical theory of communication. Bell System Technical Journal. 1948;**27**(3):379-423

[14] Alimi I, Shahpari A, Sousa A, Ferreira R, Monteiro P, Teixeira A. Challenges and opportunities of optical wireless communication technologies. In: Optical Communication Technology. London: IntechOpen; 2017. p. 41

[15] Richardson DJ, Fini JM, Nelson LE. Space-division multiplexing in optical fibres. Nature Photonics. 2013;**7**(5):354-362

[16] Yuan H, Furdek M, Muhammad A, Saljoghei A, Wosinska L, Zervas G. Space-division multiplexing in data center networks: On multi-Core fiber solutions and crosstalk-suppressed resource allocation. Journal of Optical Communications and Networking. 2018;**10**(4):272

[17] Morita I, Igarashi K, Tsuritani T. 1 Exabit/s·km transmission with multi-core fiber and spectral efficient modulation format. In: 2014 OptoElectronics and Communication Conference and Australian Conference on Optical Fibre Technology. 2014. pp. 316-318

[18] Pinho C et al. Spatial light modulator based flexible coupling platform for applications in SDM and PIC. In: 20th European Conference on Integrated Optics (ECIO 2018). 2018. pp. 238-240

[19] Richardson DJ. Unleashing the spatial domain in optical fiber communications. In: 2013 IEEE Photonics Society Summer Topical Meeting Series. 2013. pp. 70-71

[20] Essiambre R-J, Tkach RW. Capacity trends and limits of optical communication networks. Proceedings of the IEEE. 2012;**100**(5):1035-1055

[21] Carpenter J, Leon-Saval S, Eggleton BJ, Schröder J. Spatial light modulators for sub-systems and characterization in SDM. In: 2014 OptoElectronics and Communication Conference and Australian Conference on Optical Fibre Technology. 2014. pp. 23-24

[22] Lee HJ, Moon HS, Choi S-K, Park HS. Multi-core fiber interferometer using spatial light modulators for measurement of the inter-core group index differences. Optics Express. 2015;**23**(10):12555

[23] Pinho C et al. Flexible spatial light modulator based coupling platform for photonic integrated processors. International Journal on Advances in Telecommunications. 2018;**11**(1):20-31

[24] Smit M et al. An introduction to InP-based generic integration technology. Semiconductor Science and Technology—IOPscience. 2014;**29**(83001):1-41

[25] Goodman JW. Introduction to Fourier Optics. 2nd ed. Stanford: McGraw-Hill Companies; 1996

[26] Carpenter J. Holographic Mode Division Multiplexing in Optical Fibres. Cambridge, UK: University of Cambridge; 2012

[27] Lazarev G, Hermerschmidt A, Kr S. LCOS spatial light modulators: Trends and applications. In: Optical Imaging and Metrology: Advanced Technologies. Weinheim: Wiley-VCH. 2012. pp. 1-29

[28] Holoeye. "Spatial Light Modulators," Holoeye Photonics AG. 2013. [Online]. Available: http://holoeye.com/spatial-light-modulators/ [Accessed: 22-11-2017]

[29] Hamamatsu. Phase spatial light modulator LCOS-SLM. In: Opto-Semiconductor Handbook. Tokyo: Hamamatsu Photonics. 2012. pp. 1-14

[30] Kovachev M et al. Reconstruction of computer generated holograms by spatial light modulators. In: Multimedia Content Representation, Classification and Security. Vol. 4105. Berlin Heidelberg: Springer; 2006. pp. 706-713

[31] Younus SH, Hussein AT, Alresheedi MT, Elmirghani JMH. CGH for indoor visible light communication system. IEEE Access. 2017;**5**:24988-25004

[32] Torii Y, Balladares-Ocana L, Martinez-Castro J. An iterative Fourier transform algorithm for digital hologram generation using phase-only information and its implementation in a fixed-point digital signal processor. Optik (Stuttgart). 2013;**124**(22):5416-5421

[33] Ripoll O, Kettunen V, Herzig HP. Review of iterative Fourier-transform algorithms for beam shaping applications. Optical Engineering. 2004;**43**(11):2549-2556

[34] Pinho C, Lima M, Teixeira A. Optical compensation approach for SDM systems. In: XIII Symposium on Enabling Optical Networks and Sensors (SEONs 2016). 2016

[35] Lesem LB, Hirsch PM, Jordan JA. The Kinoform: A new wavefront reconstruction device. IBM Journal of Research and Development. 1969;**13**(2):150-155

[36] Carpenter J, Wilkinson TD. Graphics processing unit–accelerated holography by simulated annealing. Optical Engineering. 2010;**49**(9):095801-095807

[37] Gerchberg R, Saxton WO, Gerchberg BRW, Saxton WO. A practical algorithm for the determination of phase from image and diffraction plane pictures. Optik (Stuttgart). 1972;**35**(2):237-246

[38] Collings N, Davey T, Christmas J, Chu D, Crossland B. The applications and technology of phase-only liquid crystal on silicon devices. Journal of Display Technology. 2011;**7**(3):112-119

[39] Roelens MAF et al. Dispersion trimming in a reconfigurable wavelength selective switch. Journal of Lightwave Technology. 2008;**26**(1):73-78

[40] Wang M et al. LCoS SLM study and its application in wavelength selective switch. Photonics. 2017;**4**(2):22

[41] Turunen J, Wyrowski F. Diffractive Optics for Industrial and Commercial Applications. Berlin: John Wiley & Sons; 1997

[42] Osten W, Kohler C, Liesener J. Evaluation and application of spatial light modulators for optical metrology. In: Óptica Pura y Applicada, Reunión Española de Optoelectrónica, OPTOEL'05. 2005. pp. 71-81

[43] Varga JJM et al. Preparing arbitrary pure states of spatial qudits with a single phase-only spatial light modulator. Journal of Physics Conference Series. Apr. 2015;**605**(1):012035-012036

[44] Schroder J, Roelens MAF, Du LB, Lowery AJ, Eggleton BJ. LCOS based waveshaper technology for optical signal processing and performance monitoring. In: 17th Opto-Electronics and Communications Conference, 2012. 2012. pp. 859-860

[45] Keyworth BP. ROADM subsystems and technologies. In: OFC/NFOEC Technical Digest. Optical Fiber Communication Conference, 2005. Vol. 3. 2005. p. 4

[46] de Hennin S, Wall P, Moffat SH, Keyworth BP, Colbourne PD. Addressing manufacturability and reliability of MEMS-based WSS. In: OFC/NFOEC 2007—Conference on Optical Fiber Communication and the National Fiber Optic Engineers Conference; 2007; pp. OWV1-3

[47] Strasser TA, Wagener JL. Wavelength-selective switches for ROADM applications. IEEE Journal of Selected Topics in Quantum Electronics. 2010;**16**(5):1150-1157

[48] Shiraiwa M, Furukawa H, Miyazawa T, Awaji Y, Wada N. High-speed wavelength resource reconfiguration system concurrently establishing/removing multiwavelength signals. IEEE Photonics Journal. 2016;**8**(2):1-7

[49] Lhermite J, Suran E, Kermene V, Louradour F, Desfarges-Berthelemot A, Barthélémy A. Coherent combining of 49 laser beams from a multiple core optical fiber by a spatial light modulator. Optics Express. 2010;**18**(5):4783-4789

[50] Slinger C, Cameron C, Stanley M. Computer-generated holography as a generic display technology. IEEE Computer Society. 2005;**38**(8):46-53

[51] Tricoles G. Computer generated holograms: An historical review. Applied Optics. 1987;**26**(20):4351

[52] MathWorks. R2018b—MATLAB and Simulink Product Families. The MathWorks, Inc. [Online]. Available: https://www.mathworks.com/products/new_products/latest_features.html [Accessed: 13-04-2019]

[53] Pinho C, Neto B, Morgado TM, Neto H, Lima M, Teixeira A. InP AAC for Data Compression Applications. IET Optoelectronics. 2019;**13**(2):67-71

[54] Chang F, Onohara K, Mizuochi T. Forward error correction for 100 G transport networks. IEEE Communications Magazine. 2010;**48**(3):S48-S55

[55] Matiss A et al. Performance of an integrated coherent receiver module for up to 160G DP-QPSK transmission systems. Journal of Lightwave Technology. 2011;**29**(7):1026-1032

[56] Ashok V, Balakumaran T, Gowrishankar C, Vennila ILA, Kumar AN. The fast Haar wavelet transform for signal & image processing. International Journal of Computer Science and Information Security. 2010;**7**(1):126-130

[57] Parca G, Teixeira P, Teixeira A. All-optical image processing and compression based on Haar wavelet transform. Applied Optics. 2013;**52**(12):2932-2939

Section 3

Convergence of Wireless-Optical Networks

Chapter 8

Enabling Optical Wired and Wireless Technologies for 5G and Beyond Networks

Isiaka A. Alimi, Ana Tavares, Cátia Pinho, Abdelgader M. Abdalla, Paulo P. Monteiro and António L. Teixeira

Abstract

The emerging fifth-generation mobile communications are envisaged to support massive number of deployment scenarios based on the respective use case requirements. The requirements can be efficiently attended with ultradense small-cell cloud radio access network (C-RAN) approach. However, the C-RAN architecture imposes stringent requirements on the transport networks. This book chapter presents high-capacity and low-latency optical wired and wireless networking solutions that are capable of attending to the network demands. Meanwhile, with optical communication evolutions, there has been advent of enhanced photonic integrated circuits (PICs). The PICs are capable of offering advantages such as low-power consumption, high-mechanical stability, low footprint, small dimension, enhanced functionalities, and ease of complex system architectures. Consequently, we exploit the PICs capabilities in designing and developing the physical layer architecture of the second standard of the next-generation passive optical network (NG-PON2) system. Apart from being capable of alleviating the associated losses of the transceiver, the proposed architectures aid in increasing the system power budget. Moreover, its implementation can significantly help in reducing the optical-electrical-optical conversions issue and the required number of optical connections, which are part of the main problems being faced in the miniaturization of network elements. Additionally, we present simulation results for the model validation.

Keywords: 5G, backhaul, centralized unit (CU), common public radio interface (CPRI), distributed unit (DU), fiber to the X (FTTX), fronthaul, functional split, optical wireless communication (OWC), passive optical network (PON), photonic integrated circuits (PICs), radio access network (RAN), radio over fiber (RoF)

1. Introduction

There have been growing concerns regarding the increasing number of unprecedented bandwidth-intensive mobile applications and services being experienced by the Internet. A notable cause of the increase in the traffic and the subsequent pressure on the network is the Internet of things (IoT) technologies. For instance, massive IoT (mIoT) schemes have caused remarkable revolutions in the amount of mobile devices and applications in the networks. This is in an effort to enhance the user experience in delivering enhanced mobile broadband (eMBB) services and providing ultra-reliable low-latency communication (uRLLC) for critical communication and control services. In theory, IoT comprises universal existence of a collection of things like mobile PCs, tablets, smartphones, actuators, sensors, wireless routers, as well as radio-frequency identification (RFID) tags. It is remarkable that these devices are capable of cooperating not only with each other but also with their neighbors. By this approach, they are able to achieve common network goals by means of unique addressing scheme [1, 2]. Furthermore, it has been predicted that massive number of mobile devices on which various bandwidth-intensive applications and services will be operating and will be Internet connected [3]. In actual fact, there is a tremendous demand for effective systems that are capable of delivering various services in a cost-effective manner while meeting the essential network demands. Consequently, in an effort to accomplish the next-generation mobile network technical demands, there have been intensive researches on viable solutions that can satisfy the network requirements.

Additionally, to support the anticipated massive devices, there has been general consensus that the fifth-generation (5G) wireless communication system is the viable and promising solution. Meanwhile, massive multiple-input multiple-output (M-MIMO) antenna and millimeter-wave (mm-wave) technologies are anticipated to be integrated into the 5G networks, so as to enhance the wireless system bandwidth. This is due to the fact that radio-frequency (RF)-based wireless system transmission speeds are highly constrained by the regulated RF spectrum. This limitation can be attributed to numerous advanced wireless systems and standards such as UWB (IEEE 802.15), iBurst (IEEE 802.20), WiMAX (IEEE 802.16), Wi-Fi (IEEE 802.11), as well as the cellular-based 3G and 4G. On the other hand, there is a vast amount of unexploited and underutilized frequency at high bands [2, 4] as expatiated in Section 2. Nevertheless, the radio propagation at higher frequency bands is comparatively demanding. Consequently, advanced scheme like beamforming (BF) technique is essential for radio operation at the bands. The technique will help in compensating mm-wave band inherent path loss in the radio access network (RAN) [5–7].

In addition, owing to several innovative technologies that have been implemented in the optical communications, significant improvements have been noted in the network performance [8]. Among the remarkable improvements are the increase in the network reach, optical system capacity, and the number of users that can be effectively supported. This is as a result of cutting-edge optical fiber-based technologies. The optical schemes have been increasingly advancing deeper into different access networks, in order to provide various services such as mobile backhaul/fronthaul and multitenant fiber to the X (FTTX) with some variants of fiber-based broadband network architectures as discussed in Section 3. For instance, the optical broadband network architectures, such as fiber to the curb or cabinet (FTTC), fiber to the node (FTTN), fiber to the building (FTTB), fiber to the premise (FTTP), and fiber to the home (FTTH), proffer commercial solutions to the communication network performance bottleneck, by progressively delivering services in close proximity to the numerous subscribers [2].

It is noteworthy that various 5G use cases like uRLLC and eMBB can be effectively achieved by radio elements and BSs that are not far-off the end users or wireless devices. This is due to the fact that close proximity helps in facilitating better signal quality, with lower latency and higher data rates in the system [9]. This can be effectively realized by means of passive optical network (PON) technologies such as gigabit PON (GPON), 10Gbps PON (XG-PON), as well as Ethernet PON (EPON). It is noteworthy that one of the key issues is the process of supporting different service demands with the intention of realizing ubiquitous and elastic connections. As a result, optical and wireless networks convergence is very indispensable. This is not only a cost-effective approach but also enables high-network penetration, in order to achieve the envisaged ubiquitous feature of the next-generation network (NGN) [2]. Based on this, there is a growing consensus of opinion that high-capacity optical fronthaul scheme is one of potential solutions for addressing the network demands. For instance, if the CPRI standard is to be directly employed for the transportation of a considerable number of long-term evolution-advanced (LTE-A) and/or 5G radio signals, an enormous aggregate bandwidth will be required on the backhaul/fronthaul networks [10].

Furthermore, it has been observed that the reference system architectures for the 5G standardizations are based on the notion of heterogeneous networks where mm-wave small cells are overlaid on the larger macrocells [9]. This will enable the RAN to handle the growing traffic demands. In addition, to contain the massive deployment of small-cell BSs, cloud RAN (C-RAN) has been adopted as a promising architecture to ensure effective scalability regarding deployment cost as well as energy consumption [11–14]. The C-RAN offers an innovative architecture that is really different from the traditional distributed RAN (DRAN). In the C-RAN architecture, the baseband unit (BBU) is shifted away from the cell sites where it is normally located in the DRAN. Consequently, BBU collections that are usually referred to as BBU pools are centralized at the central office (CO). With this configuration, the remote radio heads (RRHs) are left at the cell sites.

As a result, C-RAN implementation offers significant benefits such as improved system spectral efficiency and better flexibility for further RRH deployments than the DRAN. Likewise, with the centralized BBUs, C-RAN supports greener infrastructure, enhanced interference mitigation/coordination, better resource pooling, improved BS virtualization, as well as simplified management and operation. Besides, multiple technologies can be supported with smooth and scalable evolution. Furthermore, in the C-RAN architecture, the BBU pools are connected via the fronthaul network to the RRHs. It is remarkable that the de facto air interface standard that is usually employed for connecting the BBU pools to the RRHs is the common public radio interface (CPRI) protocol. This is an interface that helps in the digital baseband sample distribution on the C-RAN fronthaul. However, stringent requirements concerning jitter, latency, and the bandwidth are imposed on the fronthaul network for seamless connectivity. This makes the CPRI-based fronthaul links to be prone to flexibility and bandwidth limitations, which may prevent them from being visible solutions for the next-generation networks [11, 12]. Meanwhile, it has been noted that the 5G systems will impose higher requirements on the transport network regarding latency, bandwidth, reliability, connectivity, and software-defined networking (SDN) capability openness [15]. A number of approaches such as cooperative radio resource allocation and data compression technologies have been adopted to address the challenges; however, the fronthaul capacity demand is still considerable high [11, 12].

The viable means of addressing the capacity requirement is through the implementation of passive optical network (PON) solutions such as wavelength division multiplexed PON (WDM-PON) and ultradense WDM-PON (UDWDM-PON). The PON architectures are compatible with the 5G networks and are capable of supporting both wired and wireless services. Based on the PON architecture, individual RRH has the chance to communicate with the BBU pools using a dedicated wavelength. Besides, in the upstream direction, the aggregate wavelengths can be further multiplexed into a single shared fiber infrastructure at the remote node (RN). They can eventually be de-multiplexed at the CO [11, 12]. As aforementioned and as depicted in **Figure 1**, optical and wireless network convergence is a

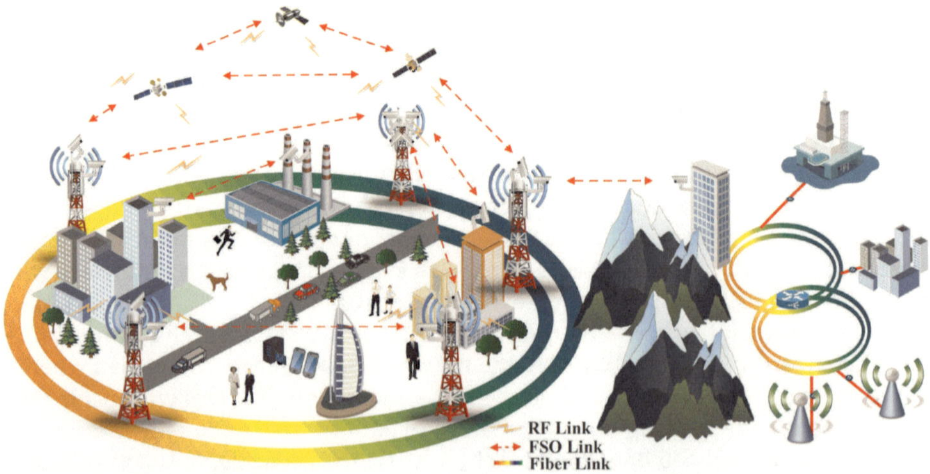

Figure 1.
A scenario for optical and wireless access networks convergence (adapted from Alimi et al. [2]).

promising scheme for exploiting the optical system inherent bandwidth and the mobility advantage of wireless connectivity, which can help in realizing the 5G network envisaged capacity and energy efficiency. In addition, optical wireless communication (OWC) is another feasible and attractive optical broadband access solution that is capable of supporting high-capacity, high-density, and low-latency networks. Therefore, it can effectively address the network requirements for different applications and services at a comparatively lower cost. So, it has been seen as an alternative and/or complementary solution for the existing wireless RF solutions [4, 16–18]. This chapter presents optical wired and wireless networking solutions for high-capacity, high-density, and low-latency networks. Furthermore, because of its potential for intense revolution and salient advantages, we focused on the second standard of the next-generation PON (NG-PON2) system. In addition, with the exploitation of notable features of photonic integration, we design and develop the physical (PHY) layer architecture of the NG-PON2 system. The proposed NG-PON2 architectures offer an enabling platform for active device integration into the chip to ensure a significantly low propagation loss. We also present simulation results for model validation. This helps in demonstrating the potential of photonic integration for optical architectures.

Furthermore, with concise information on the enabling optical wired and wireless technologies and the need for alleviating the stringent requirements in the network being introduced, we present comprehensive overview of the fronthaul transport solutions in Section 2. The salient needs for PON in the envisaged ultradense network deployments are considered in Section 3. In Section 4, a practical method for network investment optimization by the operators based on PON system coexistence is discussed. In Section 5, we present a number of viable schemes for alleviating the imposed stringent requirements in the system. The NG-PON2 PHY architecture design and development based on photonic integration are demonstrated in Section 6. In Section 7, the obtained simulation results with further discussion are presented. Section 8 concludes the chapter.

2. Fronthaul transport solutions

The fronthaul protocol can be transported by different viable means. Apart from the usually employed small form pluggable and serial constant bit rate CPRI specification that is based on digital radio over fiber (D-RoF) implementation, there are other innovative and standard fronthaul interfaces such as Open Base Station Architecture Initiative (OBSAI), next-generation fronthaul interface (NGFI), open radio interface (ORI), and enhanced CPRI (eCPRI) that can be used [19–21]. In [11], we give an overview of various prospective and standard fronthaul interfaces. In this chapter, for reference purposes, we focus on the extensively employed CPRI protocol. However, it should be noted that the transport methods to be discussed in this section are applicable to other fronthaul interfaces. The transport methods discussed in this section are grouped into wired and wireless fronthaul solutions.

2.1 Wireless fronthaul solution

Wireless transport schemes are very viable fronthaul solutions that have resulted into tremendous evolutions in the communication systems. This is due partly to the inherent advantages such as operational simplicity, ease of deployment, scalability, roaming support, effective collaboration, and cost-effectiveness. Furthermore, it is an appropriate scheme for complementing fiber-based fronthaul solutions. However, their susceptibility to transmission channel conditions makes their implementation effective for short range. Besides, the current solution can only support few CPRI interface options. This brings about bandwidth limitation for this solution. Moreover, to alleviate this, promising wireless technologies like mm-wave and wireless fidelity (Wi-Fi) can be employed in the fronthaul [11, 22, 23].

As aforementioned in Section 1, there is a huge amount of unexploited and underutilized frequency at high bands. The fronthaul in which mm-wave is being employed is feasible due to the availability of various compact and high-dimensional antenna arrays for commercial use in the band. Besides, as a result of 60 GHz standards like 802.11ad, 802.15.3c, and WirelessHD that have been issued, considerable attention has been given to mm-wave communications. Nonetheless, the inherent high propagation losses of the mm-wave communications give rise to comparatively shorter transmission range [11, 22, 24, 25].

In addition, as stated in Section 1, RF-based system transmission speeds are substantially limited due to a number of advanced wireless systems being deployed in the network. Consequently, to meet the demands of the current and future wireless networks, many chipset suppliers and wireless operators have been paying significant attention to the unlicensed spectrum. The major focus is in the 2.4 GHz and 5 GHz frequency bands that are under implementation by the Wi-Fi. This is being used for the 5G LTE-Unlicensed communication systems [11, 26]. With this implementation, the unlicensed spectrum resources could be effectively allotted to the LTE system, in order to have more capacity for supporting the Wi-Fi users [27].

Furthermore, it is remarkable that the Wi-Fi unlicensed spectrum is a promising solution for the fronthaul network. A notable advantage of exploiting the unlicensed spectrum for the fronthaul network is due to the fact that separate frequency procurement for the fronthaul might not be necessary for the network providers. Besides, the same spectrum could be effectively reused in the access and fronthaul links. This can be accomplished by means of time-division multiplexing (TDM) and frequency-division multiplexing (FDM) schemes. Another way of achieving this is through opportunistic fronthauling, in which unlicensed spectrum can be sensed. For instance, the RRH can sense unlicensed spectrum that is available (unused unlicensed spectrum) and then employ it for fronthauling. Besides, in a situation where the active user signal is considerably lower than the predefined threshold, the RRH can also make use of the spectrum. In addition, the fronthaul link constraints could be effectively eased via the Wi-Fi. This is majorly due to the fact that it can be employed for offloading [26]. Although Wi-Fi networks are capable of offering relatively high-data rates, they exhibit limited mobility and coverage. The drawbacks can be reduced by employing Wi-Fi mesh networks [11, 28].

2.2 Wired fronthaul solution

The wired network offers a number of advantages such as low interference, enhanced coverage, low latency, and high reliability and security. Due to these advantages, they have been able to stand the test of time and continue to be relevant despite the advent of wireless systems. Some of the fronthaul solutions that are based on wired links are dark fiber, passive WDM, WDM-PON, WDM/optical transport network (OTN), and Ethernet. In this subsection, we present potential wired-based fronthaul solutions that can support the network requirements.

2.2.1 Dark fiber solution

Dark fiber offers an attractive fronthaul solution. With this implementation, transmission equipment is not required between the BBU pools and the radio remote units (RRUs), consequently resulting in easiest deployment solution with least possible latency. Nevertheless, since dark fiber solution is based on point-to-point (P2P) direct connections, it lacks the required network protection, making it not a good candidate to support 5G use cases such as uRLLC services in which high reliability is required. Besides, its implementation demands huge amount of fiber resources. In the 5G systems in which ultradense networks are envisaged, the required amount of fiber is even more challenging. So, the fiber resources may be inadequate to support mIoT devices and other envisaged multimedia devices. Therefore, availability of fiber and the associated deployment cost may be the limiting factors for the dark fiber solution employment. This inefficiency can be addressed with the aids of different WDM and Ethernet solutions [11, 22, 23, 29].

2.2.2 Ethernet solution

In Ethernet-based fronthaul solution, packet technologies that encourage statistical multiplexing feature are employed. This helps in achieving traffic convergence and in enhancing the line bandwidth usage. Besides, considerable fiber resources can be saved due to its support for point-to-multipoint (P2M) transmission. Nevertheless, a number of issues such as identification as well as fast forwarding of low latency services deserve considerable research attention in this approach. Also, further efforts are required for backward compatibility with CPRI transmission and high-precision synchronization. Based on these, the Institute of Electrical and Electronics Engineers (IEEE) has established a task group known as time-sensitive networking (TSN) which is a part of the IEEE 802.1 working group, to study the latency-sensitive Ethernet forwarding technology. Reasoning along the same lines, the IEEE 1914 next-generation fronthaul interface (NGFI) working group has been established not only for the development of the NGFI transport architectures and the associated requirements but also for the definition of radio signal encapsulation specification into Ethernet packets [11, 29].

2.2.3 WDM-based solution

The requirement for low-latency transmission in the range of 10-Gb/s makes WDM-based network the usually adopted option for the fronthaul links. At large, WDM-based fronthaul methods can be grouped into two solutions which are active and passive. In active solution, other protocols are used for the CPRI traffic encapsulation, before being multiplexed on the fronthaul network. Also, the solution offers robust network topologies with considerable flexibility. Moreover, with optical amplifiers, the network reach can be significantly extended. Another important distinguishing feature of an active solution is that the cell site demarcation point requires power supply for operation. On the other hand, a passive solution mainly depends on CPRI link passive multiplexing (MUX)/demultiplexing (DEMUX). Besides, this solution's demarcation point can function effectively without any battery backup and power supply. Nonetheless, active equipment can be employed for the system monitoring at the CO demarcation point [11, 22, 23, 29].

In general, the main dissimilarities between the passive and active solutions can be recognized in the nature of their routing table and switching granularity. For instance, unlike the active solution, routing table can be statically and dynamically configured as well as associated with the interface; that of passive solution is fixed and lacks configuration capability. Likewise, the passive solution switching granularity is based on spectrum or time slot as being implemented in the TWDM-PON, while an active solution presents finer switching granularity which can be based on packet or frame switching. Consequently, the active solution offers better configuration flexibility; however, it is power-consuming and relatively complicated [12]. In the following, we expatiate on different WDM-based fronthaul solutions.

2.2.3.1 Passive WDM

In this approach, a passive optical MUX/DEMUX is employed for multiplexing a number of wavelengths on a shared optical fiber infrastructure for onward transmission. Therefore, the implementation can save considerable fiber resources via the support for multiple channels per fiber. Also, the employed optical components introduce negligible latency, so, the stipulated jitter and latency requirements for CPRI transport can be effectively met. Moreover, due to the passive nature, power supply is not required for the associated equipment operation. This brings about high power efficiency in the network. Besides, this approach is not only a cost-effective solution but also offers simple maintenance. Nevertheless, the cost implication of the wireless equipment deserves significant attention. This is due to the required colored optical interfaces at the BBU and RRU. Also, factors that need consideration are the limited transmission range and inadequate optical power budget of a relatively complex topology such as chain or ring network. This can be attributed to the accumulated insertion loss owing to multiple passive WDM components. Besides, the approach offers no robust operations, administration, and maintenance (OAM) potentials, and usually, line protection is not provided. Passive WDM implementation can also be limited by the need for well-defined network demarcation points [11, 22, 23, 29].

2.1.3.2 WDM/OTN

When WDM/OTN scheme is employed, multiplexed and transparent signal transmissions can be achieved over the fronthaul link to multiple sites. Thus, the fiber capacity is increased by enabling multiple channels on a shared fiber infrastructure [11, 23, 29]. This can be realized by encapsulating the inphase and quadrature component (I/Q) data by means of OTN frame; this is subsequently multiplexed to the WDM wavelength. Consequently, any wavelength can be employed for routing the resulting frame to the destination port [12]. Apart from being able to save fiber resources, other notable advantages of this solution are provision for OAM capabilities, network protection, service reliability, as well as service level agreement (SLA) management and network demarcation. Furthermore, this solution presents attractive features regarding low latency and high bandwidth. It is also a good approach for attending to the required colored optical interface at BBU and RRU by the passive WDM. Since colored optical interface is not demanded, wireless equipment deployment challenges are alleviated drastically by the WDM/OTN solution. Another significant advantage of the approach is the offered easy scalability. This is due to the fact that there is no need for replacing the wireless equipment optical interfaces while upgrading from non-C-RAN to the C-RAN architecture. Notwithstanding, the major drawback of the solution is the relatively higher cost of the equipment. Although power supply is not required for WDM transport in the approach, it is essential for wavelength translation and active management [11, 23, 29].

In addition, the WDM-based systems such as coarse WDM (CWDM) and dense WDM (DWDM) exhibit promising features for the fronthaul transport applications. For instance, apart from the offered high throughput and low latency, CWDM is very cost-effective regarding fiber resource usage and equipment expenses. Also, DWDM is widely known for the higher channel counts that can be efficiently supported. This can help further in increasing the number of small cells and the associated RRHs that can be deployed effectively. Furthermore, it helps in improving the fiber resource efficiency.

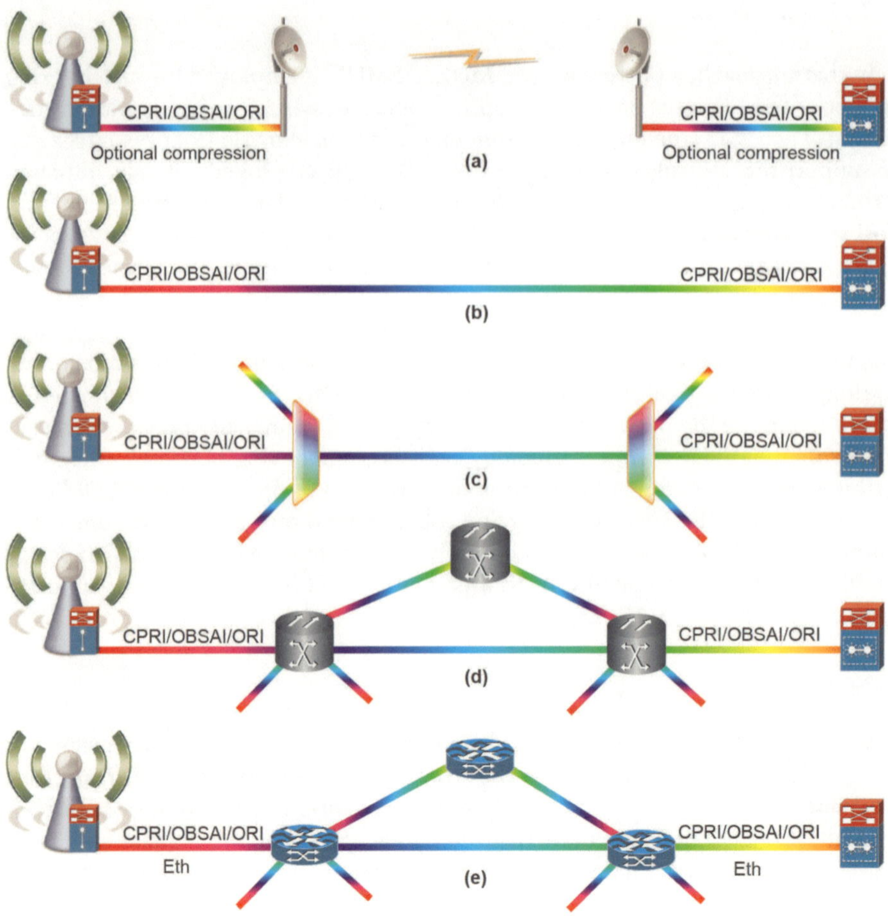

Figure 2.
Potential 5G fronthaul solutions: (a) microwave, (b) point-to-point, (c) WDM-PON, (d) OTN, and (e) Ethernet.

It is remarkable that WDM-based schemes can be used in conjunction with PON technology in order to further enhance the system performance. This scheme is highly appropriate for the anticipated massive RRHs and ultradense small cell deployment as explicated in Section 3. It should be noted that, for RAN to be well deployed, especially in the urban environments, the radio elements should be, as much as possible, in close proximity to the subscribers. So, the remote elements could be mounted on different places such as buildings and street lamp poles. Therefore, the arbitrary nature of the remote element placement can be efficiently supported with the implementation of WDM schemes.

Furthermore, as discussed, there are a number of ways by which the C-RAN fronthaul can be realized; nonetheless, the imposed stringent requirements make fiber-based method the widely adopted in the C-RAN. However, optical fiber implementation for ultradense networks, besides being time-consuming, may render the C-RAN schemes uneconomical and less flexible. It is remarkable that wireless fronthaul offers attractive and flexible solutions for information exchange between the centralized unit (CU) and distributed unit (DU). This is owing majorly to the offered advantages such as higher flexibility, lower cost, and undemanding deployment when than the fixed wired fronthaul counterparts. Therefore, innovative optical wireless solutions with good scalability and operational simplicity, coupled with easy of deployment, are really desirable [11].

In addition, apart from physical fiber-based methods being discussed, OWC system, also known as a free-space optical (FSO) communication system, is another attractive and feasible optical wireless fronthaul. The FSO provides a range of benefits such as low latency and high capacity that make it viable for addressing network requirements in a cost-effective manner [4, 16–18]. The potentials for the FSO implementation in the fronthaul network and different innovative concepts that are appropriate for improving the FSO system performance, while easing the stringent system requirements, are discussed in Section 5. Different potential 5G fronthaul solutions are depicted in **Figure 2**.

3. Passive optical network (PON)

The existing fiber-based methods as well as active P2P Ethernet might unable to meet the envisaged bandwidth-intensive traffic requirements by the 5G and beyond networks. For instance, ultradense network deployments with the associated huge network resources are envisaged in the 5G network. As illustrated in **Figure 3**, PON system can make better use of the current fiber infrastructures than the existing P2P system such as CPRI. This helps considerably in reducing the required number of interfaces in the network. As a result, it aids not only in reducing the site space, but also substantial amount of system power can be saved [30]. As explained in Section 2, PON technology has been deemed as an attractive access network solution owing to the presented advantages such as low-operation cost, high bandwidth, and low-maintenance cost [11, 31, 32].

It should be noted that the PON architectures have been experiencing continuous and gradual evolution, so as to considerably enhance the service availability and the related data rates. The offered technological options and the intrinsic benefits have been attracting the operators in deploying a number of PON systems. It has been observed that the most widely deployed one is the gigabit PON (GPON) system. Moreover, the first standard 10 Gbps PON technology, the next-generation PON (NG-PON) system, known as 10-gigabit PON (XG-PON1) has also been gaining considerable attention. With continuous demand for further capacity, there are innovative PON generations such as 10-gigabit symmetric PON (XGS-PON)

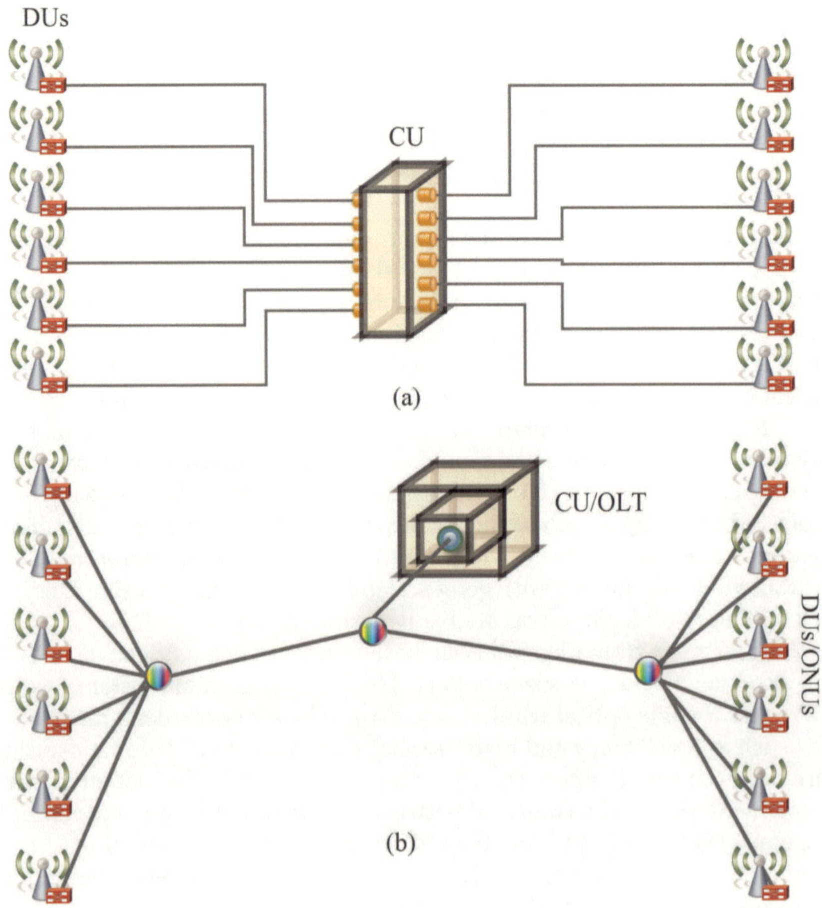

Figure 3.
Potential fronthaul solutions (a) CPRI-based and (b) PON-based schemes.

and the second standard of NG-PON (NG-PON2) that are now becoming the target of various providers [33]. In PON system, WDM and TDM techniques are normally employed to further enhance the capacity and fiber efficiency. Based on these techniques, the PON system can be broadly grouped into WDM-PON and TDM-PON.

Moreover, it is noteworthy that the TDM-PON is capable of giving considerable greater bandwidth for various data applications; however, availability of the resources that can be delivered to the end users is limited. In contrast, the issue can be effectively addressed with the WDM-PON scheme. This can be done by assigning a peculiar wavelength per subscriber. As a result of this, a distinct, high-data rate, as well as secure P2P channel, can be delivered over a high-capacity and longer optical reach, between each of the subscriber and the CU. Consequently, a WDM-PON scheme is suitable for partitioning the ONUs into a number of distinctive virtual P2P links over the shared physical optical infrastructure by multiple operators. This attribute facilitates fiber efficiency compared to P2P Ethernet. Similarly, in relation to TDM-based systems, it gives lower latency. These features make WDM-PON a disruptive solution that is very appropriate for FTTX as well as mobile backhaul and fronthaul applications. This will eventually aid the operators not only in developing converged networks but also in enhancing the current access networks. As a consequence of this, some redundant COs can be eliminated in an attempt to enhance the network performance in cost-effective ways [11, 31, 32].

Moreover, it is remarkable that advantages of both WDM-PON and TDM-PON can be effectively exploited though joint application of the schemes. This results in the TWDM-PON architecture. The potential PON architectures and their applications in telecommunication systems are presented in the subsequent subsections.

3.1 TDM-PON application

The TDM-PON can be grouped into broadband PON (BPON), asynchronous transfer mode (ATM) PON (APON), Ethernet PON (EPON), and GPON. In the existing telecommunication networks, GPON and EPON are the widely adopted schemes. Therefore, in the following, we focus on both schemes.

3.1.1 EPON application

The data traffic being encapsulated in the Ethernet frames as defined by the IEEE 802.3 standard is transported by the EPON solution. Different network elements such as optical network unit (ONU), optical line terminal (OLT), and optical distribution network (ODN) are the building blocks of a standard EPON system and other PON architectures. In the EPON solution, PON topology is exploited for getting the Ethernet access. Based on the joint schemes, EPON solution is capable of offering high bandwidth and good network scalability. Besides, due to the fact that it is highly compatible with Ethernet, network management can be supported in cost-effective manners. Likewise, as illustrated in **Figure 4**, FTTB, FTTC, and FTTH network architectures can be supported depending on the ONU deployments and demarcation point between the copper cable and optical fiber termination [32].

Typically, ONUs can be deployed beside the telegraph pole junction boxes, or else, at roadside when FTTC system is employed. Also, different types of twisted pair cables can be utilized for connecting the ONUs and the respective customer. It has been observed that FTTC technology offers a cost-effective and practical solution for delivering narrowband services. However, FTTC solution is not an ideal scheme, when broadband and narrowband services are to be incorporated [32].

Moreover, the ONU deployment can be made closer to the users in the FTTB solution. So, it can be located inside the buildings through further optical fiber penetration into customer homes. This can be achieved by means of cables, local

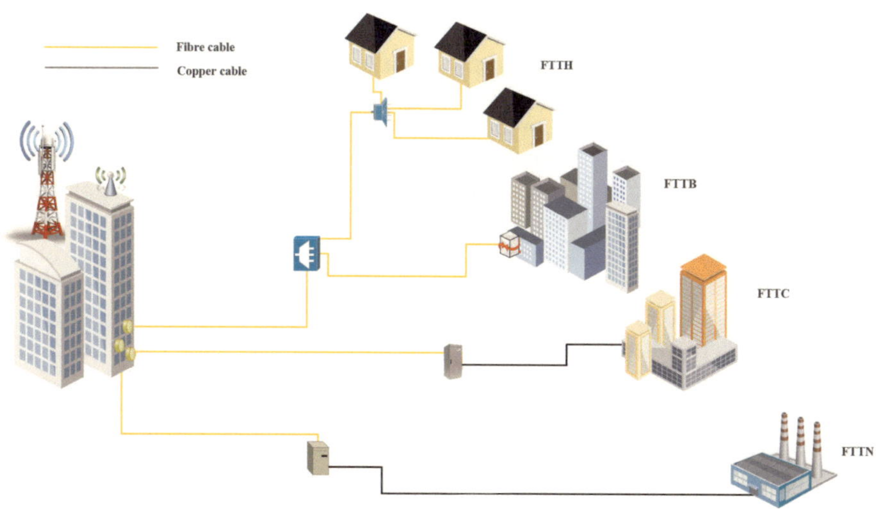

Figure 4.
FTTX architectures.

area networks (LANs), or asymmetric digital subscriber line (ADSL) broadband communication technologies. Relatively, FTTB employs more optical fiber in the connection than FTTC solution. This makes it more appropriate for broadband/narrowband service integration [32].

Furthermore, ONU deployment can take place right inside the subscribers' homes or offices in the FTTH solution. This facilitates a fully transparent network in which the ONUs are independent of the wavelength, bandwidth, as well as transmission mode and technology. These benefits enable FTTH scheme to be very ideal for access network implementations [32].

In addition, the discussed IEEE 802.3 Ethernet is a 1-Gbit/sec EPON standard. It is remarkable that there is a 10G EPON standard that is capable of supporting 10G/10G symmetric DS and US transmission. In another effort to attend to the system requirement, the IEEE 802.3ca task force has been working relentlessly on the development of 25G/50G/100G EPON standards. A notable feature of the entire EPON standards is that they are designed to be both backward and forward compatible. This is to ensure that legacy service, as well as innovative higher-speed service, can be effectively supported using the same ODN [34].

3.1.2 GPON application

Furthermore, to address the growing traffic demands, XG-PON1 has been presented. The XG-PON1 is capable of delivering higher data transmission than the legacy GPON system. Moreover, in an effort to keep the existing investments, it is backward compatible with the GPON. Also, the GPON ODN, as well as framing and management, is inherited by the XG-PON1. This encourages the reuse of the existing network elements [35].

3.2 WDM-PON application

The WDM-PON enables multiple-wavelength transmission through the multiple operators' shared optical fiber infrastructure rather than one wavelength in the PON system. This helps in ensuring that WDM-PON meets the huge subscribers' bandwidth demands. Furthermore, it presents various merits such as high wavelength efficiency and relatively simpler network management. This encourages support for various services than the TDM-PON. Besides, all anticipated services can be delivered over a shared communication network infrastructure.

In addition, it can effectively support different access networks such as FTTB, FTTH, and FTTC. Also, both small-scale and large-scale subscribers can be concurrently supported as well. Based on the inherent huge bandwidth, different types of BS bandwidth requirements can be appropriately met. Its implementation can also help in the network reach extension and in the current EPON network transition. This will help in keeping the current network investment while enhancing the network scalability [32]. In addition, UDWDM-PON offers a wavelength grid that is relatively denser for the WDM scheme. This helps not only in supporting a huge amount of aggregated wavelengths per fiber but also in accommodating higher number of RRHs per feeder fiber. Nonetheless, with the envisaged NGN stringent transport network requirements, UDWDM will be unable to maintain the high per-wavelength bit rates resourcefully. For instance, subcarriers' aggregation for high-speed services usually bring about considerable latency. Therefore, UDWDM implementation is preferred in situations where there are ultradense RRH deployments and inadequate feeder fiber accessibility. Besides, it also finds application in antenna sites which demand a low-peak but high sustainable rate [6]. As discussed in subsection 3.3, WDM-PON can be employed along with TDM-PON to achieve a hybrid WDM-TDM-PON solution known as time and wavelength division multiplexed (TWDM-PON) scheme. Apart from being efficient for both small-scale and large-scale subscribers, the hybrid scheme offers a promising solution for applications in telecommunication environment.

3.3 TWDM-PON application

It is notable that TDM-PON implementation in the 4G networks offers a very cost-efficient solution for a wavelength channel sharing between the cell sites, by means of diverse time slot allocations for different cell sites. However, with the evolution of mobile networks, the major ITU-defined application scenarios such as eMBB, uRLLC, and massive machine-type communications (mMTC) could make TDM-PON solution unsuitable for the fronthaul transport network in the 5G and beyond networks. As aforementioned, a hybrid TWDM-PON scheme is a feasible solution with abundant bandwidth capable of supporting the fronthaul demands.

With the scheme, time slots, as well as wavelength resources, can be allocated dynamically between the RRHs. The offered centralized and virtualized PON BS can considerably help in the system energy savings. Likewise, the virtualized scheme presents a number of advantages such as low handover delay, excessive handover reduction, and better network reliability. This results in cost saving, cell-edge user throughput improvement, and enhanced mobility management [32, 36, 37]. The associated multiple wavelengths, as well as potential for wavelength tenability, give TWDM-PON unprecedented means of improving the network functionalities compared with the basic TDM-PONs [36, 37]. Likewise, orthogonal frequency-division multiplexed PON (OFDM-PON) is another promising PON solution. With OFDM, there is a comparable high potential for flexible bandwidth resource sharing as experienced in the TWDM. On the other hand, regarding the reach, the OFDM variants in which direct detection is employed usually present poor performance. Similarly, variants in which coherent detection is implemented are comparatively too expensive [6]. Furthermore, it is noteworthy that among its counterparts such as standard WDM-PON, optical code division multiplexed PON (OCDM-PON), and OFDM-PON that are capable of offering 40 Gb/s or higher (80 Gb/s) aggregated bandwidth, the full service access network (FSAN) community has chosen TWDM-PON as a major broadband solution. Apart from the inherent huge capacity with 1:64 splitting ratio, it has a long reach of 40 km. The salient features enable TWDM-PON system to meet the future broadband service requirements [37–39].

A typical TWDM-PON system architecture is depicted in **Figure 5**. In a conventional TWDM-PON solution, multiple wavelengths can effectively coexist in a shared ODN by means of WDM. Moreover, each of the wavelengths is capable of serving multiple ONUs through TDM access. With reference to the ITU-T recommendation, 4–8 wavelengths in L band (1590–1610 nm) and C band (1520–1540 nm) can be employed for the downstream (DS) and upstream (US) transmissions, respectively. Also, each of the DS wavelengths can operate at 10 Gb/s, while the US can function each at 2.5 or 10 Gb/s data rate [32, 37].

In addition, the TWDM-PON ONUs employ colorless tunable transceivers for selective transmission/reception of any US/DS wavelengths (data) via a pair of US/DS wavelengths. With this approach, the ONU inventory issue can be prevented. In essence, the transceiver features help in easing network deployment as well as inventory management. Furthermore, load balancing can be supported effectively in the TWDM-PON system. Besides, with dynamic wavelength and bandwidth allocation (DWBA) implementation, large bandwidth can be flexibly exploited. It is remarkable that TWDM-PON is a stack of four 10-gigabit PONs (XG-PONs) with

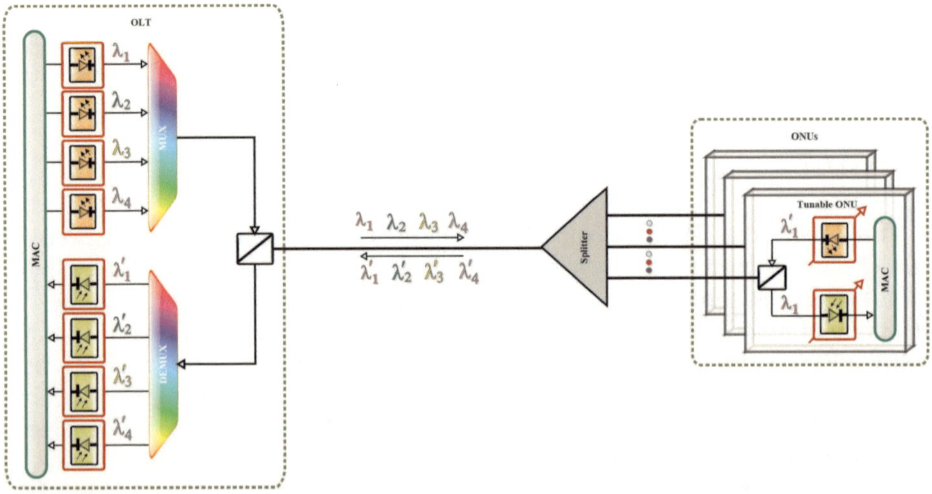

Figure 5.
Typical TWDM-PON architecture.

four pairs of wavelengths. In the stack, each XG-PON is operating on different wavelengths. Also, as stated earlier, the GPON and XG-PON GEM frames are compatible with and can be employed in the TWDM-PON solution. Based on this and the ability for coexistence with existing PON solutions, it is a viable scheme for optical access network swift evolution [11, 32, 37]. Consequently, TWDM-PON has been adopted for the NG-PON2. In NG-PON2, TWDM-PON can be employed with optional P2P WDM overlay extension. It is remarkable that DWDM scheme will enable NG-PON2 to deliver multiple unshared P2P connections, while TDM scheme simultaneously offers multiple P2M connections. This will enable the operators to efficiently support both fronthaul/backhaul and business services with the P2P WDM overlay technology, by using dedicated wavelengths [11, 40, 41].

In addition, based on the inherent colorless tunable transceivers of the TWDM-PON ONUs, three classes of wavelength channel tuning time have been specified for the NG-PON2 by the physical media dependent layer recommendation (ITU-T G.989.2). **Table 1** illustrates the specified tuning time classes by the G.989.2 recommendation. It should be noted that different innovative technologies can be exploited by the wavelength tunable devices in order to have the capability for supporting various classes. This will enable a number of potentials for the NG-PON2 system at relatively different costs. Out of the defined three classes, Class 3 is based on the slowest tunable devices. Consequently, it is applicable in scenarios with occasional tuning operations or in applications that can tolerate short service disruption. On the other hand, Class 1 wavelength tunable devices present the shortest tuning time. This feature makes them attractive for offering DWBA feature in the network. Besides, with this class implementation, the ONU transmission wavelengths can be dynamically controlled by the OLT for wavelength hopping between the transmission periods [42].

	Class		
	1	**2**	**3**
Tuning time	<10 μs	10 μs to 25 ms	25 ms to 1 s

Table 1.
Tuning time classes [42].

Although a TWDM-PON offers effective bandwidth resource allocation among multiple clients, meeting the low latency and jitter requirements of certain services may be challenging. Consequently, its implementation for the NGN RAN transport network depends mainly on the RAN use cases and deployment scenario requirements [6]. In Section 5, we present a number of viable means for alleviating the growing stringent requirements in the system. Furthermore, as aforementioned, the NG-PON2 system employs multiple wavelengths that demand for tunable transceivers at the ONUs. However, this requirement might hinder its implementation as the existing optical tunable transceivers are uneconomical. Based on this, a number of operators have been looking for ways around this by envisaging provisional scheme adoption before the full NG-PON2 migration. This will enable them to have a seamless transition with least possible or no disruption in the offered services. One of viable solution is the XGS-PON. It offers an improved commercial solution as a result of the less costly elements being employed.

3.4 XGS-PON application

The XGS-PON presents a novel technology that offers a generic solution for the NG-PON system. It can be viewed as an uncomplicated variant of TWDM-PON in which the wavelength tunability and mobility are eliminated for a more cost-effective reason. In addition, there can be an efficient coexistence between the XGS-PON and TWDM-PON using the same fiber infrastructure, since the employed wavelengths by each technology are different. Consequently, the operators can exploit the lower-cost XGS-PON for quick delivery of 10 Gbps services. This will also enable them to seize 10 Gbps services opportunities for immediate deployments. With XGS-PON, there can be cost-efficient, gradual upgrade, and well-controlled transition to a full TWDM-PON system, with minimum or no disruption to the offered services. It can also facilitate TWDM-PON system by enabling its deployment using the wavelength by wavelength approach. This will really help in pay-as-you-grow scheme for effective system upgrade and migration [33, 43].

Besides its capability for delivering 10 Gbps in both US and DS directions, XGS-PON has high potential for the dual rate transmission support as well [44]. Based on this, the 10/2.5G XG-PON ONUs and 10/10G XGS-PON ONUs can be coupled to the same OLT port via a native dual US rate TDMA scheme. It is remarkable that XGS-PON dual rate presents a comparable cost to XG-PON; nonetheless, it is capable of providing 4 times of the XG-PON US bandwidth. In addition, XGS-PON has been seen as a transitional scheme to NG-PON2 due to its ability for offering the associated NG-PON2 high-data rates in conjunction with the XG-PON1 CAPEX efficiency [33, 43]. Furthermore, it should be noted that the GPON employs 1490 and 1310 nm in the DS and UP, respectively. Likewise, XGS-PON utilizes 1578 and 1270 nm in the DS and UP, respectively. This implies that the XGS-PON service can be effectively overlaid on the same infrastructure as that of GPON. Similarly, the G.989 standard is employed in NG-PON2. The G.989 supports TWDM technologies and it is a multiwavelength access standard [44].

In addition, NG-PON2 is not only a state-of-the-art PON technology with the potential for intense revolution in the operational models of providers but also offers them flexible platform that is capable of enhancing their agility to the market demands as never before. Besides, it has the ability for cost-effective support for both the scale and capacity of the existing gigabit services while at the same time having more than enough room for the multi-gigabit bandwidth requirements of the future networks [38]. Consequently, based on the aforementioned advantages and its proficiency for multiple networks converging with outstanding performance, in this work, we focus on the NG-PON2 system. Its PHY architecture and development are presented in Section 6.

4. PON system coexistence

Furthermore, in an effort to make considerable profit, different operators have been developing high-bandwidth demanding applications and services. Good examples of such notable ultra-broadband systems are high-definition television (TV) and mIoT. It has been envisaged that there will be a further increase in the bandwidth demand due to the innovative services such as online gaming, home video editing, interactive e-learning, next-generation 3D TV, and remote medical services. However, it should be noted that NG-PON system deployment entails huge initial investments. For instance, in the greenfield FTTH systems, out of the total network investments, the ODN deployment takes between 70 and 76%. Therefore, network investment optimization can be achieved by the operators with the existing ODN exploitation. Besides, compatibility between the NG-PON evolution and the present GPON system is highly essential [35, 44].

Moreover, efficient support for bandwidth-intensive applications and services depends on coexistence of different PON technologies. The coexistence will help in the network investment optimization when the existing ODNs are shared. For instance, a network in which service delivery is being offered by GPON and needs upgrade in order to support new FTTH access technologies can coexist with the PON technologies such as XGS-PON and NG-PON2. This can be realized with the aids of a coexistence element. Based on the desired scenario, various ONT and OLT

		Class							
		A	**B**	**B+**	**C**	**N1**	**N2**	**E1**	**E2**
Loss	Min. (dB)	5	10	13	15	14	16	18	20
	Max. (dB)	20	25	28	30	29	31	33	35

Note: The degree of severity of specific class requirements could vary from one system category to another.

Table 2.
ODN optical path loss classes [42, 46].

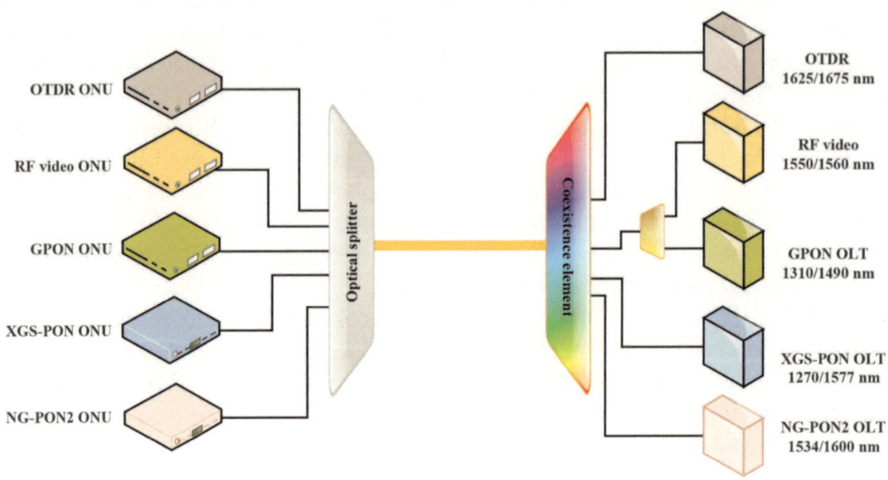

Figure 6.
PON system coexistence.

generations can effectively coexist over a shared ODN fiber infrastructure. Besides, optical time-domain reflectometer (OTDR) and RF signals can also coexist with the PON systems. This is mainly due to the fact that there is no wavelength overlap between each of the technologies. So, this permits in-band measurement without any service interruption [34, 45]. Different ODN optical path loss classes are presented in **Table 2**.

It is remarkable that, apart from the fact that the existing GPON subscribers can be kept together with higher-bandwidth services, the coexistence will also give the operators the profound chance to take advantage of different approaches such as asymmetrical and symmetrical data rates. They also have deployment flexibility by operating on fixed or tunable wavelengths in order to offer appropriate operations and services at suitable costs. It will also assist the operators in the NG-PON evolution path not only by allowing them to upgrade their networks accordingly but also for gradual migration to the evolving PON technologies that are capable of offering the full optical potential. Thus, they have the liberty of adopting the cost and deployment pace that best fit their precise business requirements [43]. Moreover, this will enable the operators in making further revenue by exploiting flexible bandwidth and wavelength plans in order to support any service type as well as any business need. **Figure 6** depicts a PON system coexistence for a gradual and pay-as-you-grow expansion [33].

5. System requirement alleviation schemes

As explained in Section 1, C-RAN is envisioned to be a promising candidate for efficient management of the access network and the associated emergent complexity. This is due in part to its cost-effectiveness and remarkable flexibility for the network element deployments. Normally, the inphase and quadrature (I/Q) component stream transmission in this architecture is via the D-RoF-based CPRI. It is remarkable that CPRI-based fronthaul demands huge bandwidth which could be a limiting factor in the 5G and beyond networks in which mm-wave and massive MIMO are anticipated to be implemented. Consequently, an advanced optical transmission technology such as analog RoF (ARoF) has to be employed for an efficient fronthaul solution realization [11, 13, 14].

5.1 RoF schemes

The RoF schemes offer efficient and economical methods for modulated RF signal transmission. For instance, it can be used for transmission from the CO, to a number of distributed RRHs, through low-loss optical fiber networks, by employing an optical carrier. In addition, as aforementioned in Section 1, optical and wireless network convergence is highly imperative for scalable and cost-effective broadband wireless networks. The envisaged convergence for the next-generation mobile communication networks can be efficiently achieved with the implementation of RoF. This is due to its simplicity and efficiency in conveying wireless signal via an optical carrier. Furthermore, the inherent low attenuation and huge bandwidth of optical link can effectively support multiple wireless services on a shared optical fronthaul network. Moreover, with RoF implementation, the CUs and DUs can be well-supported. This offers effective centralized network control that subsequently presents advantages such as easy upgrade, simple maintenance, and efficient resource sharing [11, 47, 48].

It should be noted that there are various RoF options that can be employed in the network. Furthermore, each of the viable options presents related distinct merits and demerits. Out of the variants, the highly spectrally efficient scheme is the ARoF. Besides, its implementation results in a most power-efficient and least complex RRH design. Nevertheless, it is susceptible to intermodulation distortion which is as a result of optical and microwave component nonlinearity. This results in relatively shorter operating distance. Moreover, the transmitter components such as oscillators, digital to analog converters (DACs), and mixers consume a considerable amount of power. On the other hand, with D-RoF implementation, the ARoF-associated nonlinearity issue can be effectively mitigated. However, in a scenario where high baud rates and high carrier frequencies are required, the DAC power consumption and expenditure are excessively high. Also, if upconversion is required or implemented at the RRHs, it turns out to be substantially high. Consequently, having a fixed phase relation among various RRHs is really challenging. Besides, digitized sample transmission, rather than the analog signal, brings about a significantly low spectral efficiency. The aforementioned drawbacks can be more challenging when densely distributed RRHs are to be supported [11, 47, 48]. Therefore, to address the challenges, a hybrid scheme that is capable of exploiting the ARoF and D-RoF schemes can be employed. One of notable techniques for a hybrid scheme is based on the implementation of sigma-delta-over-fiber (SDoF). This scheme helps in ensuring digital transmission that can support simple and power-efficient RRHs. Besides, there is no need for high-resolution and high-speed DACs with its implementation [47].

It is noteworthy that the RoF scheme employment is contingent on physical optical fiber availability. On the other hand, for the envisaged ultradense small-cell deployment, fiber deployment is not only time-consuming but also capital intensive. Likewise, there could be inappropriate system deployment due to the associated right-of-way acquisition. For these reasons, as well as limited number of the deployed fiber, the FSO system practicability has been considered [11, 13, 14].

5.2 FSO scheme

FSO communication presents an alternative technology for optical fiber systems. It is based on RF signal transmission between the CU and the DU apertures via the free space. Therefore, being an optical wireless technology, the fiber media are not required, and, consequently, trenches are unnecessary for its implementation. Moreover, like a well-developed, viable, and widely employed RoF technology, FSO scheme is capable of supporting multiple RF signal transmission. Apart from having inherent optical fiber features like RoF, FSO scheme offers additional merits regarding time-saving and cost-effectiveness, since there is no need for physical fiber deployment. This makes it to be very applicable in scenarios where physical network connectivity through optical fiber media is challenging and/or unrealistic. Besides, it is capable of delivering broadband services in rural area where there is an inadequate fiber infrastructure [11, 13, 14]. It is noteworthy that, when employed as a complementary solution for fronthauling, FSO can be a promising mobile traffic offloading scheme for alleviating the stringent requirements of bandwidth-intensive services transmission via the mobile networks.

In addition, the FSO scheme offers a number of benefits such as high bit rates, ease of deployment, full duplex transmission, license-free operation, improved protocol transparency, and high-transmission security. These salient merits enable the FSO scheme to be considered as a viable broadband access technology. It is capable of addressing various services and applications' bandwidth requirements at low cost for the NGNs. Based on these, the RF signals over FSO (RoFSO) idea have been presented. This is in an effort to exploit the inherent massive transport capacity of optical systems and the related deployment simplicity of wireless networks [11, 13, 14].

Furthermore, a DWDM RoFSO scheme implementation has the capability of supporting concurrent multiple wireless signal transmission [49]. Nevertheless, the FSO systems have some drawbacks due to their susceptibility to the atmospheric turbulence and local weather conditions. The effects of these can cause beam wandering, as well as scintillation, which in due course results in the received optical intensity fluctuation. Consequently, the system reliability and availability can be determined by the extent of the effects. As a result, FSO technology is relatively unreliable like the normal optical fiber technology. Therefore, apart from the fact that these can limit the RoFSO system performance, its employment for uRLLC applications might also be limited as well. Consequently, the drawbacks hinder the FSO scheme as an effective standalone solution. Therefore, for the FSO scheme to be effective, the associated turbulence-induced fading has to be alleviated [2, 17, 18, 50]. Based on this, several PHY layer ideas like maximum likelihood sequence detection, diversity schemes, adaptive optics, and error control coding with interleaving have been presented to address the issue [11, 50, 51]. Besides, innovative schemes such as relay-assisted transmission and hybrid RF/FSO technologies can be implemented to enhance the system performance regarding capacity, reliability, and availability [11].

5.3 Hybrid RF/FSO scheme

A hybrid RF/FSO scheme exploits the inherent high-transmission bandwidth of the optical wireless system and the related deployment simplicity of wireless links [2]. In addition, the hybrid RF/FSO system idea does not only base on concurrent means of attending to the hybrid scheme related limitations, but it also entails ways of exploiting both approaches for a reliable heterogeneous wireless service delivery. The hybrid scheme is able to achieve this by incorporating the RF solutions' scalability and cost-effectiveness with the FSO solutions' high data rate and low latency. Consequently, the technology is able to address the high throughput, cost-effectiveness, and low-latency requirements of the system. Besides, it presents a heterogeneous platform for wireless service provisioning for the envisaged 5G and beyond networks [11, 13, 14, 52, 53].

5.4 Relay-assisted FSO scheme

One of feasible methods of turbulence-induced fading mitigation is the spatial diversity scheme. In this technique, there are multiple deployed apertures at the receiver and/or transmitter sides. This is in an effort to realize extra degrees of freedom in the spatial domain. It is remarkable that spatial diversity is an appealing fading mitigation scheme, owing to the presented redundancy feature. On the other hand, multiple-aperture deployment in the system causes a number of challenges like an increase in the cost and system complexity. Moreover, in order to prevent the spatial correlation detrimental effects, the aperture separation should be sufficiently large. Furthermore, a notable approach for simplified spatial diversity implementation is a dual-hop relaying scheme. It is noteworthy that there has been extensive implementation of the scheme in the RF and wireless communication systems. Application of the scheme in these fields not only aids in improving the receive signal quality but also helps considerably in the network range extension [2, 11, 13, 14].

Conceptually, multiple virtual aperture systems are generated in the relay-assisted transmission with the intention of realizing salient MIMO technique features. The architecture takes advantage of the RF and FSO features for an efficient and reliable service delivery. In addition, a relay-assisted transmission system is an innovative communication technique known as a mixed RF/FSO dual-hop communication system. The dual-hop scheme meaning can be easily understood from its architecture. In the architecture, the transport networks from the source to the relay system are RF links; however, the transport networks between the relay system and the associated destination node(s) are FSO links. Hence, in a dual-hop system, RF is used for signal transmission at one hop, while FSO transmission is implemented at the other. The FSO link mainly functions to facilitate the RF users' communication with the backbone network. This is purposely for filling the connectivity gap between the backbone and the last-mile access networks. Accordingly, the offered architecture can efficiently address the system-related last-mile transmission bottleneck. This can be effectively achieved by supporting multiplexed users with RF capacities. The users can also be aggregated onto a shared high-capacity FSO link. This will help in harnessing the inherent huge bandwidth of an optical communication system. Another outstanding advantage of this scheme is that any kind of interference can be easily inhibited via its implementation. This is due mainly to the fact that the RF and FSO operating frequency bands are completely different. Consequently, it offers better performance than the traditional RF/RF transmission schemes [2, 11, 13, 14].

5.5 RAN functional split

The RAN functional split is another innovative and practical scheme for alleviating the imposed fronthaul requirements by the C-RAN architecture [11, 54]. For instance, to address the drawbacks of CPRI-based fronthaul solutions, an eCPRI specification presents additional physical layer functional split options and a packet-based solution. Consequently, unlike the conventional constant data rate CPRI in which the stream significantly depends on the carrier bandwidth, as well as the number of antennas, the eCPRI stream does not depend on either of the factors but on the actual traffic load. In essence, apart from being able to alleviate the stringent bandwidth demands, multiple eCPRI stream can also be multiplexed onto a wavelength for onward transmission over the fronthaul network [12].

In addition, with recent network architecture development, the traditional BBU and RRU have been reformed into different functional entities which are the CU, DU, and RRU/active antenna unit (AAU). With the configuration, the CU majorly focuses on non-real time and part of the traditional Evolved Packet Core functionalities. This involves high-level protocol processing like dual connectivity and radio resource management. In addition, the DU is responsible for the real-time media

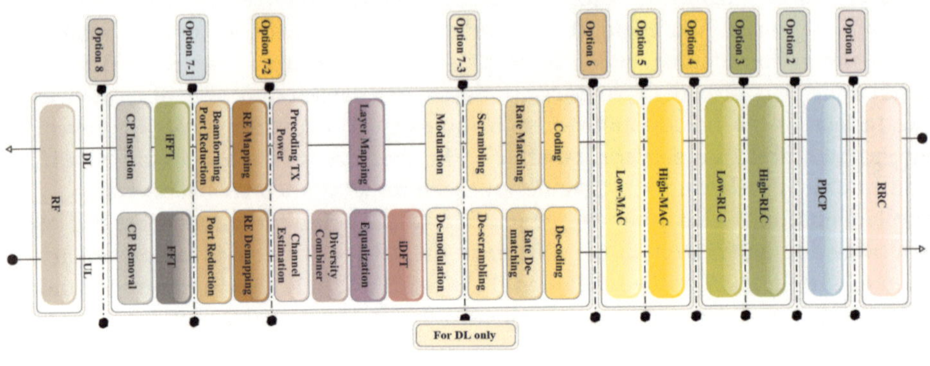

Figure 7.
Functional split options between CU and DU with emphasized PHY layer.

access control layer functions like HARQ flow and physical layer function processing. Also, when massive MIMO antennas are to be employed, certain parts of the physical layer functions can also be shifted to the RRU/AAU. The implementation will not only aid in lessening the associated transmission bandwidth between the RRU/AAU and DUs but will also help in reducing the transmission cost considerably. Therefore, a number of functional split options have been presented in order to reduce the processing and network resource cost considerably. As shown in **Figure 7**, each of the option is categorized according to the demarcation point between the CU and the DU. Therefore, depending on the deployment scenarios and use cases, each option offers different degrees of flexibility regarding resource allocation for different service requirements [12, 29].

6. NG-PON2 physical layer architecture design and development

The NG-PON2 physical layer requirements are very challenging. Besides, the requirements are even more strict than the legacy PON technologies. For instance, when compared with the GPON taken into consideration the related spectrum, GPON employs only one channel for the transmission and one for the reception, with a very wide wavelength allocation (up to 100 nm). On the other hand, in NG-PON2, there are <4 nm to accommodate four channels. Consequently, this means that the thermal control must be very precise in order to keep each channel inside the specified channel space (which is +/−20 GHz). As aforementioned, there are multiple channels in NG-PON2 transmission; therefore, the receiver must be tunable so as to work for any one of them at a particular time while others are rejected. This requirement implies that there is a need for a very tight band-pass filter too for efficient operation. Also, the tuning time classes, already presented in **Table 1** in Section 3, are likewise strict and difficult to achieve on the hardware side. Besides, one of the major related issues is the amount of the required optical-electrical-optical (OEO) conversions, which can bring about an unviable and unsustainable system [55].

6.1 Photonic integrated circuit

The optical communications evolution has initiated enhanced photonic integrated circuits (PICs) that present a cost-effective alternative to data transmission. With PIC technology implementation, a number of optical components such as modulators, lasers, amplifiers, detectors, etc. can be merged/integrated on a single chip. Consequently, it helps in optical system design simplification, system reliability enhancement, as well as significant power consumption and space reduction. In addition, there can be considerable reduction in the amount of OEO converters required for the system implementation. This subsequently results in the total network cost reduction [55]. Thus, it is anticipated to be an enabling and viable technology with immense flexibility and reconfigurability in a number of fields [56]. A PIC has numerous advantages over the traditional optical sub-assemblies (OSAs). For instance, considering the occupied volume, the PICs allow a very dense architecture in a small area, passing also by the optical losses; however, the losses in the OSAs are higher because of the internal free-space alignment between each optical component. Also, other notable advantages of the PICs compared with the OSAs are lower power consumption, lower footprint, and cost-effectiveness. Therefore, PICs have the capability of permitting flexible and high data rate solutions [39, 55].

In the following, for the system realization, we propose three different architectures: the ONU architecture, the OLT architecture, and the architecture that can perform both functions just by hardware selection. It should be noted that all of these architectures have the transmit and the receive parts.

6.1.1 NG-PON2 ONU transceiver architecture

The ONU transceiver architecture is represented in **Figure 8**. This is a very simple structure regarding the optical setup, but the electrical control is very tough, mostly because of the tunability (both on the transmitter and on the receiver). In this example, there is one tunable laser. The laser can be tuned by temperature and can be directly or externally modulated (the latter would also need a modulator after the laser). On the receiver part, there is an optical band-pass filter which has to be tunable to allow one of the downstream channels and cut the rest of the spectrum. The tunable band-pass filter is followed by an optical receiver.

6.1.2 NG-PON2 OLT transceiver architecture

As explained before, the OLT is not tunable; both transmitter and receiver should work on the same fixed wavelength pair, as depicted in **Figure 9**. Consequently, four pairs of optical devices will be needed. Since it is very difficult to encapsulate everything on the same transceiver, the solution that is being followed commercially is having four different transceivers, one for each wavelength pair, and the wavelength multiplexer (WM) device is external. This WM should, in each port, allow one wavelength pair, meaning that in each port, it should pass only one downstream and the respective upstream channel.

6.1.3 NG-PON2 OLT/ONU transmission architecture

The architectures presented in **Figures 8** and **9** are the basic ones to have functional devices for NG-PON2. But taking advantage of photonic integration, it is possible to develop a much more complex circuit with more functionalities, which is being presented next. **Figure 10** illustrates the block diagram of an architecture that can be used both as ONU and OLT. This helps in exploiting the advantage of both functionalities on a single chip. The purpose (OLT or ONU) to be served can be achieved just by hardware selection. This proposed architecture fits inside a 4 4.6 mm indium phosphide (InP) PIC. In the following subsection, we present the final design and some obtained simulation results.

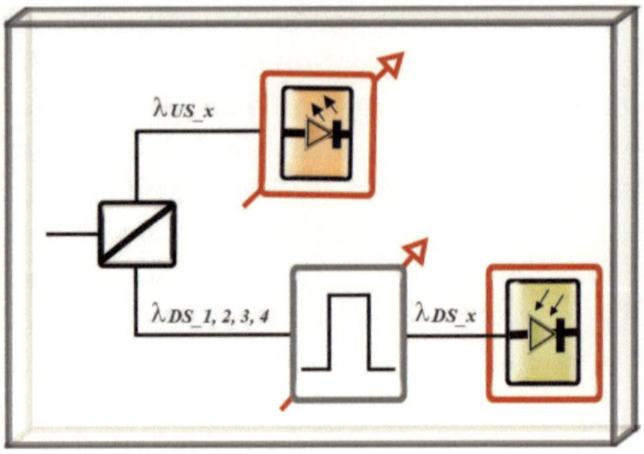

Figure 8.
ONU transceiver architecture.

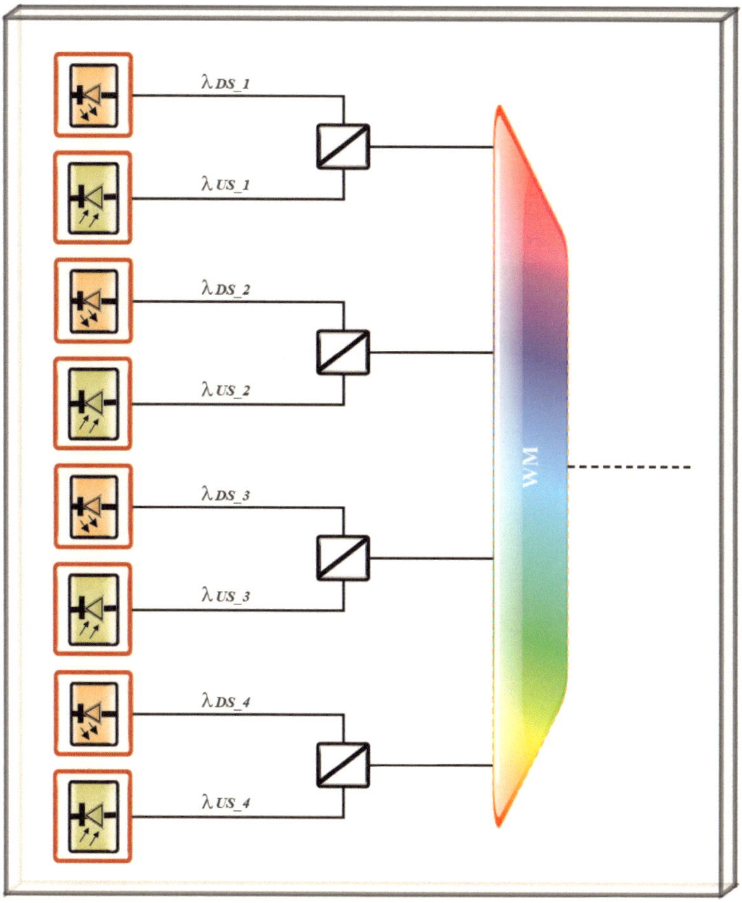

Figure 9.
OLT transceiver architecture.

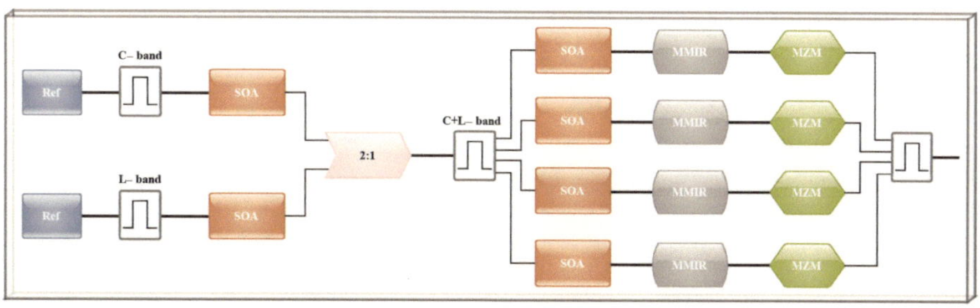

Figure 10.
Block diagram of OLT/ONU transmission architecture.

6.2 PIC implementation of OLT/ONU and receiver circuits

The architecture comprises four lasers, four Mach-Zehnder modulators (MZM), and a number of filters. Two of the filters are for changing the operational frequency band (C band for upstream transmission and L band for downstream). Also, one filter is employed for tuning the four lasers to the correct wavelength. Besides, at the output, there is one filter working as a combiner of the four lasers. The band selection is made using the two semiconductor optical amplifiers (SOAs) that are placed after the band filters. It is noteworthy that the two SOAs are working as switches and determine the chip's operating mode (i.e., OLT or ONU). Therefore, one of the SOAs is amplifying the light (active SOA), while the other is absorbing (passive SOA). Consequently, by this configuration, only one band filter is contributing to the setup. The employed lasers are built using laser cavities which contain SOAs that are being used for gain purposes, filters, and reflectors on both sides. The C + L band filter helps in the selection of the downstream or upstream channel [39].

Moreover, the architecture includes also a multimode interferometer reflector (MMIR) before the band selection and another one after each gain SOA. These reflectors define the laser cavity limits. The second MMIR, after the gain SOAs, only reflects 50% of the light, and the remaining 50% is the laser cavity output and is sent to the MZM for modulation. After the modulation on the MZMs, all four channels are combined in just one, and the resulting light signal is sent to the output

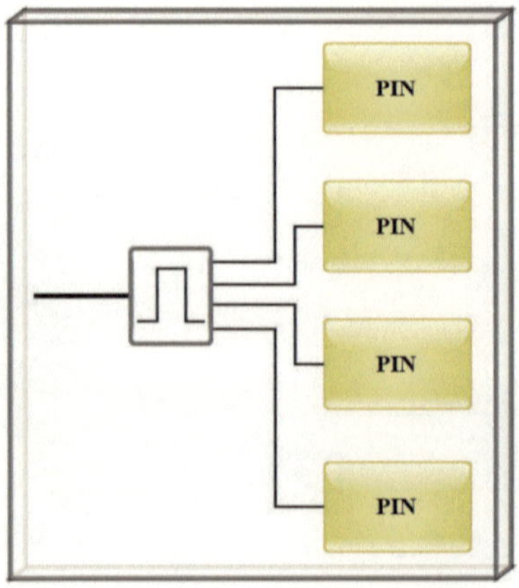

Figure 11.
Receiver block diagram.

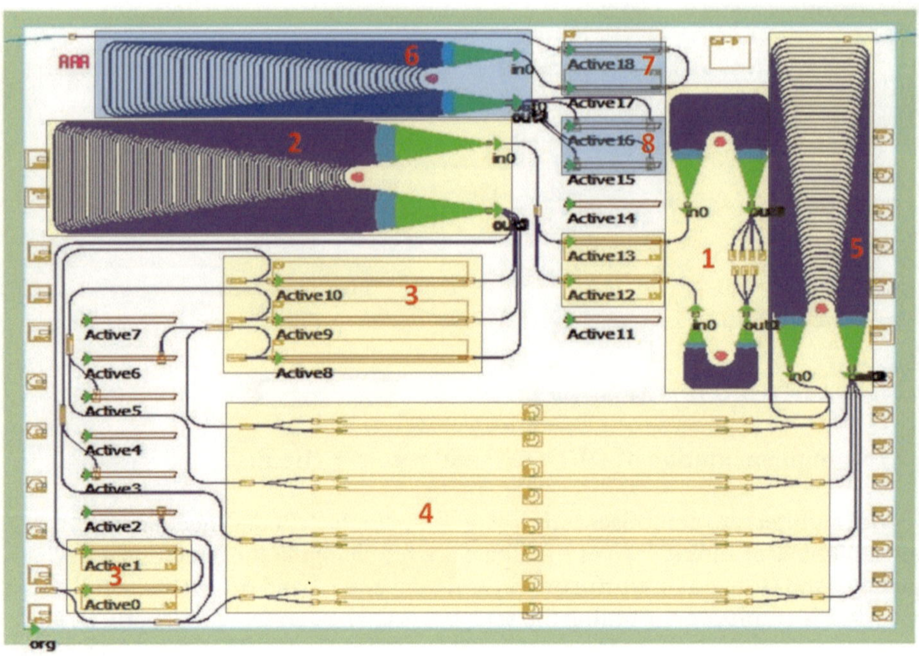

Figure 12.
OLT/ONU integrated transceiver design masks.

of the PIC, where a fiber will be aligned to collect the light, and subsequently, it will be sent to the network [39].

6.2.1 PIC implementation of receiver circuit

This PIC has also a receiver circuit, but it is a simple one, with just a wavelength division multiplexer (WDM) filter which receives the light from the network and routes each NG-PON2 channel for a different PIN. The receiver circuit schematic is depicted in **Figure 11**.

6.2.2 PIC implementation of OLT/ONU circuit

Using the photonic design kit (PDK) from the foundry Smart Photonics and a software for PIC design (Phoenix Software at the time, meantime bought by synopsis) for the implementation, the final circuit masks of the chip are shown in **Figure 12**.

7. Results and discussion

In this section, we present the obtained simulation results with further discussion on NG-PON2 physical layer architecture design and development based on PICs. **Figure 13** shows the spectral simulation results obtained using advanced simulator for photonic integrated circuits (Aspic) software from filarete. On the left figure, there is the downstream operation (L band selected), and on the right there is the upstream (C band selected). In the figure, the spectra in blue, pink, orange, and green are the four channels. In both cases, it is possible to conclude that there is about 30 dB of suppression of replicas. The suppression facilitates smooth operation of the system by preventing intra-channel interference.

The reason for using laser cavities is due to the limitations on the foundry. During the chip's design period, the Smart Photonics did not offer lasers on their process design kit (PDK). Consequently, improvements in the architecture can be undertaken to potentiate the results. For instance, the laser cavities could be replaced by distributed feedback (DFB) or distributed Bragg reflector (DBR) lasers that have narrow linewidth and a stable single mode operation. In this case, the cavity would disappear, and the filtering should be done after the lasing. In addition, the architectures can be simplified using only one modulator; nevertheless, it

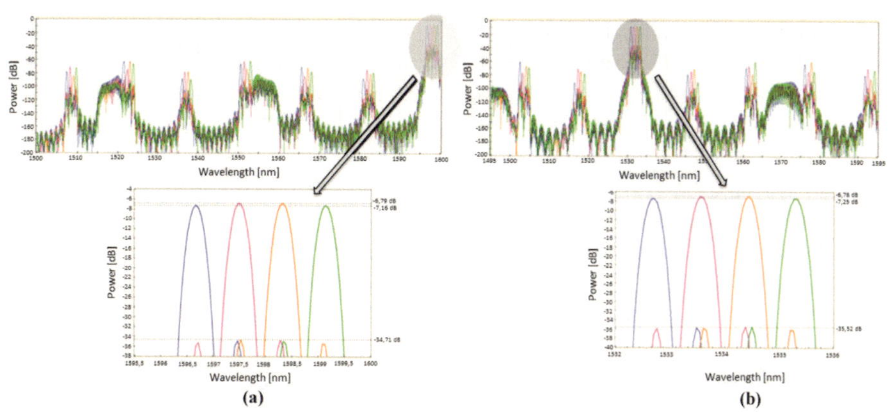

Figure 13.
Optical spectra at the transmitter output (a) downstream and (b) upstream.

would not be possible to transmit the four channels simultaneously; this implies that only one channel can be transmitted at a time. The proposed and developed architectures demonstrate the potential of photonic integration for optical architectures. Consequently, the architectures not only have the ability of supporting high data rates, high density, and flexible solutions but also offer advantages such as low power consumption, improved functionality, low footprint, and cost-effectiveness.

8. Conclusion

The 5G based system is a promising solution for attending to the growing concerns about the traffic pressure on the network. Also, the envisaged massive number of deployment scenarios and use cases to be supported brings about high-bandwidth and low-latency requirements for the 5G networks. The small-cell-based C-RAN approach can efficiently attend to the associated ultradense deployment. However, the C-RAN-based approach imposes stringent requirements regarding jitter, bandwidth, and latency for the mobile transport networks. In this book chapter, we have presented wired and wireless transport solutions that are capable of addressing the C-RAN-based stringent requirements and, consequently, the 5G mobile transport network demands. Furthermore, owing to its significant and inherent advantages for the 5G and beyond networks, we have focused on the NG-PON2 system. We have exploited the salient advantages and the low footprint platform offered by the PICs in the NG-PON2 system design and implementation. Based on these technologies, the proposed architectures are capable of alleviating the associated losses in the system while also helping in increasing the system power budget. In addition, employment of the proposed architectures can help the device makers, service/network providers, and infrastructure and chip vendors, in lowering the footprint of network elements.

Acknowledgements

This work is funded by Fundação para a Ciência e a Tecnologia (FCT) through national funds under the scholarships PD/BD/105858/2014. It is also supported by the European Regional Development Fund (FEDER), through the Regional Operational Programme of Lisbon (POR LISBOA 2020) and the Competitiveness and Internationalization Operational Programme (COMPETE 2020) of the Portugal 2020 framework, Project 5G (POCI-01-0247-FEDER-024539), ORCIP (CENTRO-01-0145-FEDER-022141), and SOCA (CENTRO-01-0145-FEDER-000010). It is also funded by Fundação para a Ciência e a Tecnologia (FCT) through national funds under the project COMPRESS-PTDC/EEI-TEL/7163/2014 and by FEDER, through the Regional Operational Program of Centre (CENTRO 2020) of the Portugal 2020 framework [Project HeatIT with Nr. 017942 (CENTRO-01-0247-FEDER-017942)] and [Project Virtual Fiber Box with Nr. 033910 (POCI-01-0247-FEDER-033910)]. Additional support is provided by the COST action CA16220 European Network for High Performance Integrated Microwave Photonics (EUIMWP) and IT (UID/EEA/50008/2013).

Author details

Isiaka A. Alimi[1*], Ana Tavares[1,2], Cátia Pinho[1], Abdelgader M. Abdalla[1], Paulo P. Monteiro[1] and António L. Teixeira[1]

1 Department of Electronics, Universidade de Aveiro, Instituto de Telecomunicações, Telecommunications and Informatics, Aveiro, Portugal

2 PICadvanced, University of Aveiro Incubator, Portugal

*Address all correspondence to: iaalimi@ua.pt

References

[1] Yu C, Yu L, Wu Y, He Y, Lu Q. Uplink scheduling and link adaptation for narrowband internet of things systems. IEEE Access. 2017;**5**:1724-1734

[2] Alimi I, Shahpari A, Sousa A, Ferreira R, Monteiro P, Teixeira A. Chapter 2: Challenges and opportunities of optical wireless communication technologies. In: Pinho P, editor. Optical Communication Technology. Rijeka: IntechOpen; 2017

[3] Ejaz W, Anpalagan A, Imran MA, Jo M, Naeem M, Qaisar SB, et al. Internet of things (iot) in 5g wireless communications. IEEE Access. 2016;**4**: 10310-10314

[4] Alimi I, Shahpari A, Ribeiro V, Kumar N, Monteiro P, Teixeira A. Optical wireless communication for future broadband access networks. In: 21st European Conference on Networks and Optical Communications (NOC); June 2016; pp. 124-128

[5] Laraqui K, Tombaz S, Furuskär A, Skubic B, Nazari A, Trojer E. Fixed Wireless Access on a Massive Scale with 5G. Ericsson Technology Review. Vol. 94. Stockholm, Sweden: Ericsson; 2017

[6] Skubic B, Fiorani M, Tombaz S, Furuskär A, Mårtensson J, Monti P. Optical transport solutions for 5G fixed wireless access [invited]. IEEE/OSA Journal of Optical Communications and Networking. 2017;**9**(9):D10-D18

[7] David Schnaufer. 5 Things to Consider When Designing Fixed Wireless Access (FWA) Systems. United States of America: Qorvo; 2018

[8] Pinho C, Shahpari A, Alimi I, Lima M, and Teixeira A. Optical transforms and cgh for sdm systems. In: 2016 18th International Conference on Transparent Optical Networks (ICTON); July 2016; pp. 1-4

[9] mmw 5G-eMBB use cases and small cell based HyperDense networks. Document, version: 197.10.01, Small Cell Forum, December. UK: Small Cell Forum; 2017

[10] Torres-Ferrera P, Straullu S, Abrate S, Gaudino R. Upstream and downstream analysis of an optical fronthaul system based on dsp-assisted channel aggregation. IEEE/OSA Journal of Optical Communications and Networking. 2017;**9**(12):1191-1201

[11] Alimi IA, Teixeira AL, Monteiro PP. Toward an efficient c-ran optical fronthaul for the future networks: A tutorial on technologies, requirements, challenges, and solutions. IEEE Communications Surveys Tutorials. 2018;**20**(1):708-769

[12] Yu H, Zhang J, Ji Y, Tornatore M. Energy-efficient dynamic lightpath adjustment in a decomposed awgr-based passive wdm fronthaul. IEEE/OSA Journal of Optical Communications and Networking. 2018;**10**(9):749-759

[13] Alimi IA, Monteiro PP, Teixeira AL. Analysis of multiuser mixed rf/fso relay networks for performance improvements in cloud computing-based radio access networks (cc-rans). Optics Communications. 2017;**402**: 653-661

[14] Alimi IA, Monteiro PP, Teixeira AL. Outage probability of multiuser mixed rf/fso relay schemes for heterogeneous cloud radio access networks (h-crans). Wireless Personal Communications. 2017;**95**(1):27-41

[15] Fuchuan Z. Challenges and trends for 5G transport. ZTE Technologies. 2018;**20**(2):19-21

[16] Ghassemlooy Z, Popoola W, Rajbhandari S. Optical Wireless Communications: System and Channel Modelling with MATLAB®. Boca Raton, London New York: CRC Press; 2012

[17] Alimi I, Shahpari A, Ribeiro V, Sousa A, Monteiro P, Teixeira A. Channel characterization and empirical model for ergodic capacity of free-space optical communication link. Optics Communications. 2017;**390**:123-129

[18] Alimi IA, Abdalla AM, Rodriguez J, Monteiro PP, Teixeira AL. Spatial interpolated lookup tables (luts) models for ergodic capacity of mimo fso systems. IEEE Photonics Technology Letters. 2017;**29**(7):583-586

[19] Transport network support of IMT-2020/5G. Technical report gstr-tn5g. Canada: ITU-T; February 2018

[20] Common Public Radio Interface: eCPRI Interface Specification. Interface specification, ecpri specification v1.1, eCPRI specification. CPRI cooperation; January 2018

[21] View on 5G Architecture. Architecture white paper version 2.0, 5G PPP architecture working group, July 2018

[22] Guiomar FP, Alimi IA, Monteiro PP, and Gameiro A. Flexible infrastructure for the development and integration of access/fronthauling solutions in future wireless systems. In: 2018 IEEE 19th International Workshop on Signal Processing Advances in Wireless Communications (SPAWC); June 2018; pp. 1-5

[23] Nokia Optical Anyhaul as an enabler of C-RAN: Accelerating the delivery of 5G networks. White paper, document code: Sr1803022985en, Nokia, March 2018

[24] Kalfas G, Pleros N, Alonso L, Verikoukis C. Network planning for 802.11ad and mt-mac 60 ghz fiber-wireless gigabit wireless local area networks over passive optical networks. IEEE/OSA Journal of Optical Communications and Networking. 2016;**8**(4):206-220

[25] Stephen RG, Zhang R. Joint millimeter-wave fronthaul and ofdma resource allocation in ultra-dense cran. IEEE Transactions on Communications. 2017;**65**(3):1411-1423

[26] Zhang H, Dong Y, Cheng J, Hossain MJ, Leung VCM. Fronthauling for 5g lte-u ultra dense cloud small cell networks. IEEE Wireless Communications. 2016;**23**(6):48-53

[27] Chen Q, Yu G, Maaref A, Li GY, Huang A. Rethinking mobile data offloading for lte in unlicensed spectrum. IEEE Transactions on Wireless Communications. 2016;**15**(7):4987-5000

[28] Aijaz A, Aghvami H, Amani M. A survey on mobile data offloading: Technical and business perspectives. IEEE Wireless Communications. 2013;**20**(2):104-112

[29] 5G-oriented Optical Transport Network Solution. White paper, ZTE Technologies, June 2017

[30] Otaka A. Flexible access system architecture: Fasa. In: NTT Tsukuba Forum 2016 Workshop Lectures. Tokyo, Japan: NTT; Vol. 15. 2017. pp. 1-7

[31] De Betou EI, Bunge C-A, Åhlfeldt H, Olson M. WDM-PON is a key component in next generation access. Nashua, USA: Lightwave; Article, Lightwave, March 2014

[32] The Application of TDM-PON and WDM-PON. Article, Fiber-Optical-Networking, May 2015

[33] Bogataj T. Safe Migration to Next-Gen Optical Broadband Access: A gradual and controlled journey to XGS-PON and NG-PON2. Kranj, Slovenia: Executive whitepaper, Iskratel, January 2018

[34] Xu Q. What the future holds for next-generation PON technologies. Nashua, USA: Article, Cabling Installation and Maintenance, October 2017

[35] Next-Generation PON Evolution. Shenzhen, China: Manual, Huawei Technologies. 2010

[36] Wang X, Wang L, Cavdar C, Tornatore M, Figueiredo GB, Chung HS, et al. Handover reduction in virtualized cloud radio access networks using twdm-pon fronthaul. IEEE/OSA Journal of Optical Communications and Networking. 2016;**8**(12):B124-B134

[37] Cheng N, Gao J, Xu C, Gao B, Liu D, Wang L, et al. Flexible twdm pon system with pluggable optical transceiver modules. Optics Express. 2014;**22**(2):2078-2091

[38] The Future of Passive Optical Networking is Here: NG-PON2. Marketing report, Broadband Forum

[39] Jesus Teixeira AL, Maia Tavares AC, Silva Lopes AP, Rodrigues CE. Photonic integrated tunable multi-wavelength transmitter circuit. USA: United States Patent; 2017

[40] Nesset D. Ng-pon2 technology and standards. Journal of Lightwave Technology. 2015;**33**(5):1136-1143

[41] Monteiro PP, Viana D, da Silva J, Riscado D, Drummond M, Oliveira ASR, et al. Mobile fronthaul rof transceivers for c-ran applications. In: 2015 17th International Conference on Transparent Optical Networks (ICTON); July 2015. pp. 1-4

[42] Wey JS, Nesset D, Valvo M, Grobe K, Roberts H, Luo Y, et al. Physical layer aspects of ng-pon2 standards—Part 1: Optical link design [invited]. IEEE/OSA Journal of Optical Communications and Networking. 2016;**8**(1):33-42

[43] Thomas D. XGS-PON makes NG-PON simpler. Technical report, Nokia, June 2016

[44] Challenges in Next-Gen PON Deployment. White paper, Viavi Solutions, 2017

[45] WaveCEx: WDM module for PON coexistence-GPON, XGS-PON, NG-PON2, RF, OTDR. Wavecex family: 180409 data sheet, TELNET Redes Inteligentes, October 2017

[46] OBA000100 GPON Fundamentals. Technical report, Huawei

[47] Breyne L, Torfs G, Yin X, Demeester P, Bauwelinck J. Comparison between analog radio-over-fiber and sigma delta modulated radio-over-fiber. IEEE Photonics Technology Letters. 2017;**29**(21):1808-1811

[48] Thomas VA, Ghafoor S, El-Hajjar M, Hanzo L. A full-duplex diversity-assisted hybrid analogue/digitized radio over fibre for optical/wireless integration. IEEE Communications Letters. 2013;**17**(2):409-412

[49] Dat PT, Bekkali A, Kazaura K, Wakamori K, Matsumoto M. A universal platform for ubiquitous wireless communications using radio over fso system. Journal of Lightwave Technology. 2010;**28**(16):2258-2267

[50] Alimi IA, Shahpari A, Monteiro PP, Teixeira AL. Effects of diversity schemes and correlated channels on owc systems performance. Journal of Modern Optics. 2017;**64**(21):2298-2305

[51] Navidpour SM, Uysal M, Kavehrad M. Ber performance of free-space

optical transmission with spatial diversity. IEEE Transactions on Wireless Communications. 2007;**6**(8): 2813-2819

[52] Dahrouj H, Douik A, Rayal F, Al-Naffouri TY, Alouini M. Cost-effective hybrid rf/fso backhaul solution for next generation wireless systems. IEEE Wireless Communications. 2015;**22**(5): 98-104

[53] Kazaura K, Wakamori K, Matsumoto M, Higashino T, Tsukamoto K, Komaki S. Rofso: A universal platform for convergence of fiber and free-space optical communication networks. IEEE Communications Magazine. 2010;**48**(2):130-137

[54] Harutyunyan D, Riggio R. Flex5g: Flexible functional split in 5g networks. IEEE Transactions on Network and Service Management. 2018;**15**(3): 961-975

[55] Tavares A, Lopes A, Rodrigues C, Mãocheia P, Mendes T, Brandão S, et al. Photonic Integrated Transmitter and Receiver for Ng-pon2. Proc. SPIE 9286, Second International Conference on Applications of Optics and Photonics, 928605 (22 August 2014)

[56] Pinho C, Gordon GSD, Neto B, Morgado TM, Rodrigues F, Tavares A, et al. Flexible spatial light modulator based coupling platform for photonic integrated processors. International Journal on Advances in Telecommunications. 2018;**11**(1):20-31

Chapter 9

Role of Optical Network in Cloud/Fog Computing

Kiran Deep Singh

Abstract

This chapter is a study of exploring the role of the optical network in the cloud/fog computing environment. With the growing network issues, unified and cost-effective computing services and efficient utilization of optical resources are required for building smart applications. Fog computing provides the foundation platform for implementing cyber-physical system (CPS) applications which require ultra-low latency. Also, the digital revolution of fog/cloud computing using optical resources has upgraded the education system by intertwined VR using the fog nodes. Presently, the current technologies face many challenges such as ultra-low delay, optimum bandwidth, and minimum energy consumption to promote virtual reality (VR)-based and electroencephalogram (EEG)-based gaming applications. Ultra-low delay, optimum bandwidth, and minimum energy consumption. Therefore, an Optical-Fog layer is introduced to provide a novel, secure, highly distributed, and ultra-dense fog computing infrastructure. Also, for optimum utilization of optical resources, a novel concept of *OpticalFogNode* is introduced that provides computation and storage capabilities at the Optical-Fog layer in the software defined networking (SDN)-based optical network. It efficiently facilitates the dynamic deployment of new distributed SDN-based *OpticalFogNode* which supports low-latency services with minimum energy as well as bandwidth usage. Therefore, an EEG-based VR framework is also introduced that uses the resources of the optical network in the cloud/fog computing environment.

Keywords: cloud/fog computing, optical resources, virtual reality, cyber physical system, electroencephalogram (EEG)

1. Introduction

Optical transmission is the most cost-effective technology to implement high-bandwidth-based communication in the fog/cloud computing environment. The passive optical network (PON) uses optical line terminals (OLT) and optical network units (ONU) for delivering fog/cloud-based services effectively [1]. Introducing this technology with the present information technology, Internet of Things (IoT), cloud computing, 5G wireless networking, and embedded artificial intelligence have tremendous potential to assist the development of smart applications that demand a large amount of data to be processed locally and operate on-premise with minimum latency and network congestion [2]. Optical technology has supported IoT-based applications for transferring massive information in a virtual frictionless fashion by using the optical network elements. It has provided new ways for various business applications to move over the latest technologies such as big data analytics, machine learning, etc. in the era of the 5G network.

Fog computing is a new distributed architecture which brings computing, storage, and networking services closer to the proximity of the end user [3]. As compared to the traditional cloud computing techniques, it processes real-time applications and data at the edge with minimum latency, minimum network congestion, and lower energy consumption which are the key demand of many industries such as manufacturing, e-health, education, oil and gas, smart cities, smart homes, and smart grids [4]. Fog nodes aggregate the computing resources of edge devices to perform the critical data-sensitive computations where the data of analysis part is directly sent to the cloud for further processing because traditional fog nodes have limited storage and computing power.

The integration of fog computing and PON is an inexpensive, scalable, and simple technology to provide a most promising solution for building e-learning-based smart educational applications [5]. The dynamic capabilities of SDN combined with the state-of-the-art optical technologies have the ability to modernize the optical transport network through its primary feature, i.e., programmability [6]. The purpose of this chapter is to explore efficient techniques to combine SDN-based optical technology at different levels of design and development of smart VR-based applications.

2. Utilization of optical resources in cloud/fog computing environment

In order to handle real-time and bandwidth-intensive applications, fog leverages the computing resources of the SDN-based optical network. **Figure 1** shows that the Optical-Fog layer [7] uses ONUs in the middleware of the cloud and IoT layer. In a typical PON channel with multiple OLTs, each OLT is connected with multiple ONUs (16–256) [8]. The Optical-Fog layer is designed by using their residual processing, storage, and interconnection capabilities. It can enable fast service provisioning, dynamic service restoration, network automation, and network optimization at different layers of the underlying network infrastructure. It makes optical network

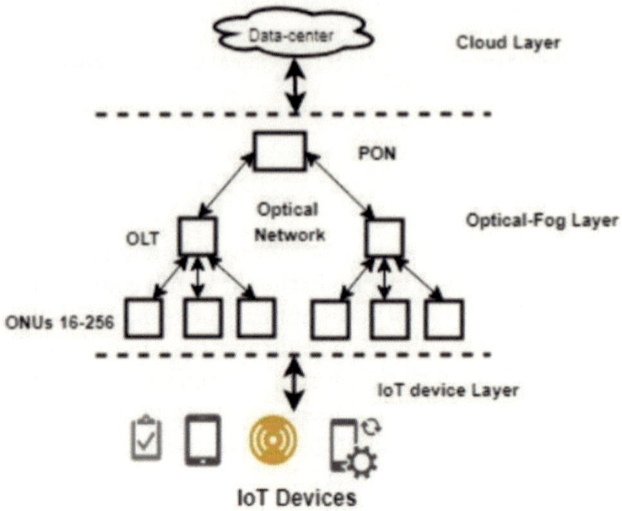

Figure 1.
Utilization of optical resources in cloud/fog computing environment.

centralized, intelligent, and controlled (real time) which can serve application-level services efficiently in the heterogeneous IoT, machine learning, big data, and cloud computing paradigms. The acronyms used in this paper are defined in **Table 1**.

2.1 SDN-based optical network

Presently, SDN is supporting wide area network to deal with many more technologies for delivering several benefits. It has adopted a hierarchical approach in which domain controllers collect information and delegate the control (real time) over the network layers and geographic clusters to support applications and provide higher levels of service orchestrations. Initially, SDN was used in data centers for separating the data plane, control plane, and management plane from each other [9]. The interface like OpenFlow is used by the centralized controller to deliver computing infrastructure for making better communication. While applying this concept to the optical network, optical domain controller (ODC) plays an important role. As shown in **Figure 2**, it provides a more programmatic and abstract view of the underlying optical network through the northbound interface [10]. The programming feature of SDN makes it capable of fulfilling customized demands for manipulating network infrastructure. To handle real-time, bandwidth-intensive applications, fog uses the computing resources of the SDN-based optical network.

The SDN-based optical network infrastructure fulfills the demand of increasingly high-performance and network-based applications with flexibility and efficiency. The key security issues in fog/cloud computing over optical network lies at both downstream and upstream channels of PON. PON uses broadcasting in the downstream channel which is prone to eavesdropping attacks where an attacker can modify the behavior of ONUs at its media access control (MAC) layer. On the other hand, the traffic in the upstream channel is only visible to the OLT rather than other ONUs that can also be exploited for attacks. In PON network, OLT uses time division multiplexing access (TDMA) that provides sharing of the upstream channel among

PON	Passive optical network
OLT	Optical line terminal
ONU	Optical network unit
CPS	Cyber physical systems
VR	Virtual reality
EEG	Electroencephalogram
QoE	Quality of experience
QoS	Quality of service
ONV	Optical network virtualization
SDN	Social-defined network
FAR	Free available resource
TDMA	Time division multiplexing access
MPCP	Multipoint control protocol
ODC	Optical domain controller
MAC	Media access control

Table 1.
List of acronyms.

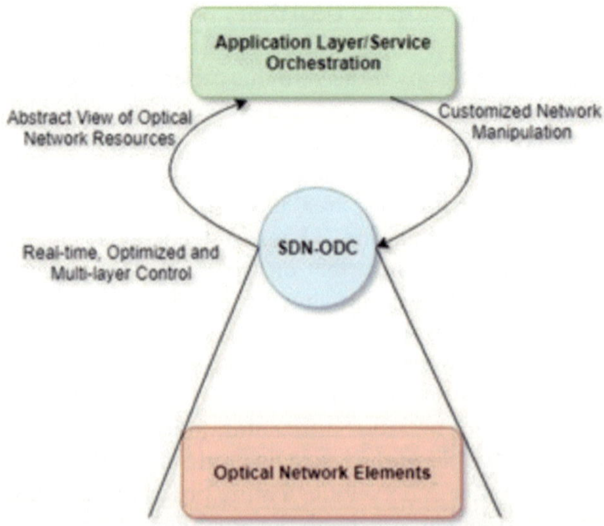

Figure 2.
SDN-based optical network.

ONUs [11]. It assigns static or dynamic nonoverlapping time slots to connected ONUs. Here, OLT and all connected ONUs are well synchronized that result in the collision-free transmission of the traffic or data frames. In the security aspects of PON, more research work is required that can restrict the ONUs to send data frames outside of their preassigned time-slots. In case if a malicious ONU intends to send data frames outside its preassigned time slot, the collision can encounter with the data-frames of other ONUs that degrade the quality of service (QoS) of the optical channel.

To ensure QoS and quality of experience (QoE) to the end-user for various real-time CPS-based applications, Optical-Fog layer is utilized. To handle real-time processing, this layer uses the optical network resources by creating *OpticalFogNode.* The optical network can effectively realize the interconnected optical resources (PON, OLTs, and ONUs) across the 5G network. It provides ultralow delay and less energy consumption for IoT devices and uses the majority of the computing resources of the optical layer rather than the cloud layer.

2.2 *OpticalFogNode* for implementing CPS system

CPS enables the integration of cyber components such as sensors, computational and control units, and network devices into the physical components (such as objects, end-users, and infrastructures) by connecting them to the Internet and each other. It also has shown tremendous progress in many fields like communication, healthcare, education, manufacturing, robotics, transportation, military, etc. It has also encouraged many innovative and ever-growing projects in the application domain of cloud and fog computing. CPS requires a novel, highly distributed, secure fog computing infrastructure in the heterogeneous network for strengthening its position for the mobile and wireless network in the new 5G era [12]. It can provide unified and cost-effective computing services for smart cities, vertical industries, and IoTs at the extreme edge of the new 5G network.

Further, the concept of an *OpticalFogNode* is proposed that supports low-cost and on-demand access to the computing infrastructure of the Optical-Fog layer in the 5G network. The main challenge is to run the CPS-based applications on the *OpticalFogNode.* The optical network virtualization (ONV) and SDN provide a novel solution to deploy *OpticalFogNode* at the edge of the network. All free available resources (FARs) of the optical elements are grouped together to form an *OpticalFogNode* with the computing capabilities like processor, memory, and bandwidth. ONV converts the free available physical resources of the optical network elements into the virtual resources as infrastructure-as-a-service (IaaS) model. Initially, each submitted task is categorized as CBS-based or non-CPS-based task on the basis of requested resources in terms of processing power, storage, bandwidth, acceptable security level, etc. The Optical-Fog manager can dynamically reconfigure the *OpticalFogNode* which provides the desired reliability and QoS for the CPS. An algorithm is proposed that identify all possible created *OpticalFogNode* on the SDN path and assign them CPS-based tasks for further processing. The non-CPS-based tasks are directly sent to the cloud layer only if the resources of *OpticalFogNode* are not free.

Algorithm Task placement algorithm for the proposed framework.

```
Data: T
while P ε Across all SDN paths do
    List RunningTasks;
    while OpticalFogNode ε P do
        List TaskToPlaced;
        while task T ε CPS do
            if All predecessors of T are in TaskToPlaced then
                Add T to TaskToPlaced;
            end
        end
    end
    while Task T ε TaskToPlaced do
        if (Resources_T^req < Resources_OpticalFogNode^Avail) then
            Allocate T on OpticalFogNode^Avail ;
            if (!OpticalFogNode^Avail) then
                Allocate T on OpticalFogNode_DC^Avail ;
            end
        else
            Allocate T on OpticalFogNode_DC^Avail ;
            if (!OpticalFogNode_DC^Avail) then
                Choose T' such that (T' > T AND T'=non-CPS) if
                 (!NULL) then
                    Allocate T' on cloud data-centers;
                    Allocate T on OpticalFogNode_DC^Avail;
                else
                    Allocate T on cloud data-centers;
                end
            end
        end
    end
end
```

In the proposed algorithm, the resources required by the new task are evaluated and then allocated on the *OpticalFogNode*. This node is scalable to provide the required computing resources dynamically by using the concept of ONV.

If the computing resources of *OpticalFogNode* are already occupied, then there are two options to execute the task on the basis of its preference. If the requested new task is a non-CPS task, it can be directly allocated to the cloud. Otherwise, CPS-centric task can be executed.

- *T* represents the requirements of submitted task for the framework along with its category as CPS-based or non-CPS-based task.
- *RunningTask* represents the already running tasks.
- *OpticalFogNode* is virtual and a dynamically configurable smart node using the concept of ONV at the Optical-Fog layer.
- *TaskToPlaced* all coming task to be allocated to the *OpticalFogNode* for further processing.
- $OpticalFogNodeNode^{Avail}$ is a free available resource at the OpticalFog.
- $OpticalFogNode_{DC}^{Avail}$ is a free available resource at the dynamically configured *OpticalFogNode.*

Further, this layer uses the SDN-based controller for optimizing the distribution of the flow among various redundant paths. In order to increase the QoS, the shortest path is chosen that minimizes the delay. In the optical network in the 5G environment, the *OpticalFogNode* has a flow table which is used to match the routing information of the received packet in the path. If there is no entry found in the flow table, the received packet is forwarded to the SDN controller for finding the shortest path so that the particular packet can be forwarded. Thus, a new entry is added (once the path is chosen) in the flow table of the *OpticalFogNode* for the coming future packets. Hence, the proposed SDN controller identifies the shortest path with the least congestion among all possible paths.

2.2.1 Architecture of OpticalFogNode

In SDN-based Optical-Fog/cloud network, the key challenge with the deployment of a fog node is to make it secure from the attackers. However, attackers are capable to create malicious programs with the ability to detect and evade their targets in distributed computing environments. **Figure 3** shows that optical network virtualization provides a novel solution to deploy *OpticalFogNode* in the middleware of IoT devices and the cloud rather than deploying at cloud data centers.

SDN technology combined with optical network virtualization allows for running the control logic of each tenant on a virtual SDN controller rather deploying and running at the cloud data centers.

The resources to the proposed *OpticalFogNode* can be provisioned on demand from geographically distributed optical elements specially ONUs. The architecture of *OpticalFogNode* is shown in **Figure 4** where it can be deployed and configured virtually. The architecture has southbound and northbound interface along with SDN controllers which belong to different tenants of *OpticalFogNode* to emulate them for different IoT applications. It has the capability to control optical network elements for processing the configuring demand of different *OpticalFogNode* tenants such as computing resources, topology, address scheme, node mapping options, etc. Hence, virtual *OpticalFogNode* can be created in the form of infrastructure-as-a-service for providing real-time control to each *OpticalFogNode* tenant over its virtual

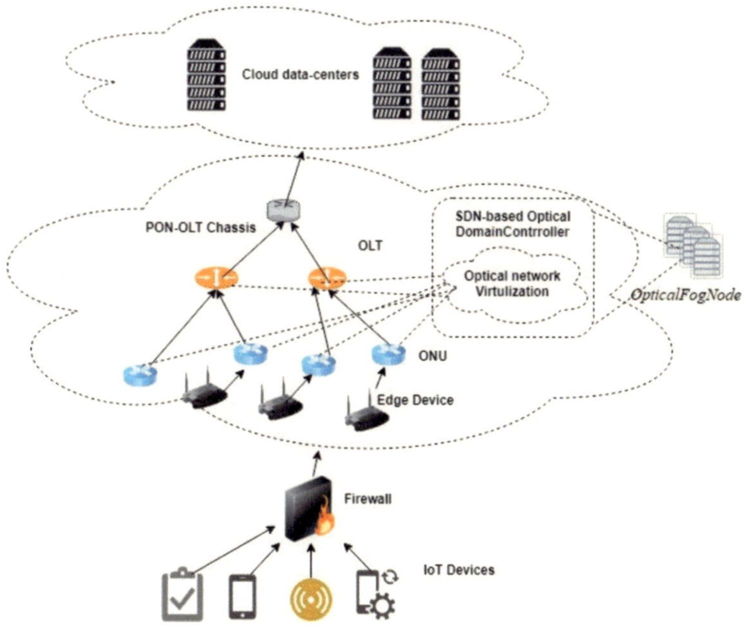

Figure 3.
Deployment scenario of OpticalFogNode at Optical-Fog layer.

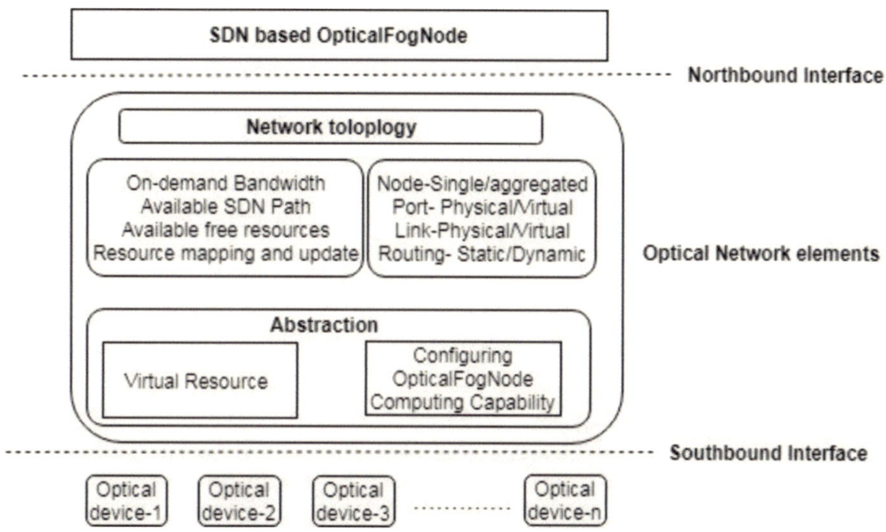

Figure 4.
SDN-based OpticalFogNode architecture.

network. In order to form virtual infrastructure, a free-available-resource concept is proposed which uses the freely available resources of the optical network elements that lie at the Optical-Fog layer. Since routers and switches have limited resources, only optical elements such as optical network units and optical line terminals are taken into account for implementing FAR. Optical elements like ONUs and OLTs have their own processing, storage, and interconnection capabilities that are not fully utilized by the present network scenario. Thus, as shown in **Figure 5**, each optical element has some amount of running resources as well as FAR. Our proposed *OpticalFogNode* aggregates those FARs for facilitating the computing capability to each *OpticalFogNode* tenant.

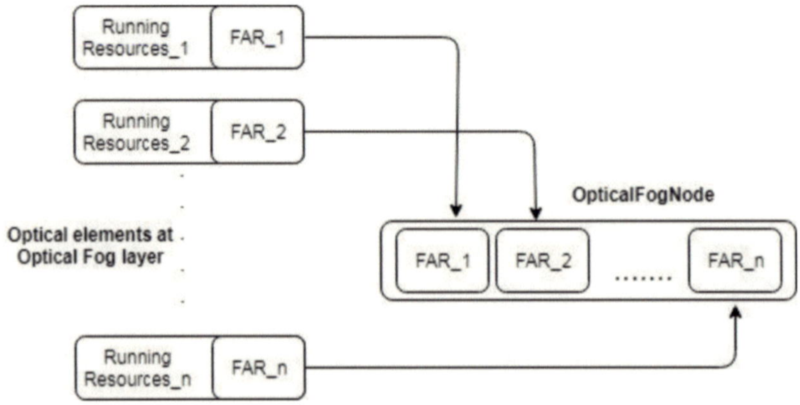

Figure 5.
Free available resource.

Hence, all FARs of optical network are grouped together to form virtual data centers with computing resources such as processor, memory, and bandwidth. ONV converts the physical resources of optical network elements into the virtual resources as infrastructure-as-a-service (IaaS) model to build virtual honeypots that prevents vulnerability and its identity from the attacker.

2.3 EEG-based VR gaming applications

SDN-based Optical-Fog network introduced as shown in **Figure 3** provides optimum bandwidth and ultralow delay for EEG-based VR gaming applications. The Edge-Fog layer and Optical-Fog layer provide rich gaming experience and QoE for EEG-based VR gaming applications by utilizing the optical resources than the cloud resources [13]. The Optical-Fog layer executes the game logic where the VR scenes can be encoded and streamed at the Edge-Fog layer. SDN-based controller improves the QoE and supports the playing of a game across the distributed geo-locations with minimum delay. It optimizes the flow distribution among the various redundant paths inside the Optical-Fog network to reduce the delay. In contrast to a traditional controller, the proposed SDN controller provides the shortest path with the least congestion among all possible paths

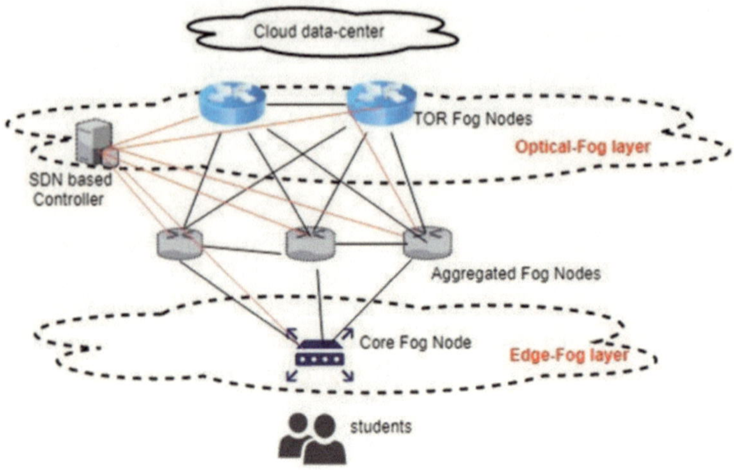

Figure 6.
SDN-based Optical-Fog network.

from the requesting Core Fog nodes to the Top of Rack (TOR) Fog nodes in the optical network which is shown in **Figure 6**. It uses an open-loop congestion control mechanism to employ congestion aware direct routing. Each node of the SDN network keeps the estimation cost $C_n^c(t)$ for delivering packets to their destination node c [14]. It helps to find the shortest path with least congestion by using the historical knowledge of the connection to node c and the waiting time of packets to c in the node n's queue. It is assumed that all nodes broadcast a request for the cost frequently to their neighbors. Also, all neighbor nodes keep updating their cost table on the basis of the received request for the cost. To find the shortest path with the least congestion, the node with minimum delivery cost is selected as shown in **Figure 7**. The convoluted parameters are referred to as *Proximity Measure* $\Theta_n^c(t)$ and *Net Destination Queue Waiting Time* $\Omega_n^c(t)$ are used to compute the delivery cost [13].

$$\Theta_n^c(t) = \frac{Q_n^c(t)}{T_n^c(t)} \tag{1}$$

The value of $\Theta_n^c(t)$ lies between 0 and 1. The value 1 indicates the connection between n and c, whereas 0 shows that they were never connected. Here, $T_n^c(t)$ is the time increment, and $Q_n^c(t)$ is the time duration while c and n remains connected.

2.3.1 Net destination queue waiting time

$$\Omega_n^c(t) = \sum_{i=0}^{N} \left(\tau - a_{n,i}^c\right) \tag{2}$$

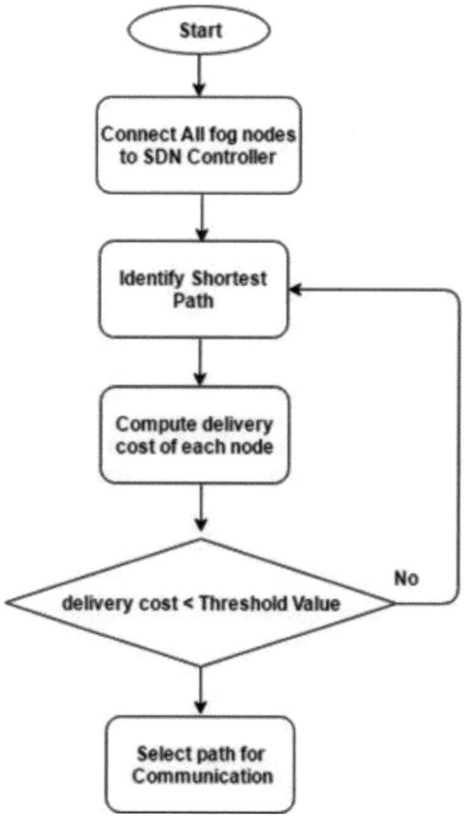

Figure 7.
Workflow to find the shortest path.

Here, τ id the present time and $a_{n,i}^{c}$ is the arrival time of packets i. Since the queue waiting time is used to predict the congestion, delivery cost can be considered as an exponentially increasing function. Hence, the delivery cost to c via n is computed as:

$$C_n^c(t) = \Omega_n^c(t).[1 - \Theta_n^c(t)] + C_n^c(t - 1) \tag{3}$$

Thus, the shortest path with least congestion is identified by pulling packets toward the neighbors that have the smallest queue. It helps the SDN controller to set the threshold value for decision making.

2.3.2 Modules placement strategy in the Optical-Fog network

In order to deploy gaming modules, an algorithm is proposed that utilizes the Edge-Fog layer and Optical-Fog layer in the Optical-Fog network. The proposed algorithm places gaming modules using the SDN topology and iterates over all paths. Here, it places modules on the devices in incremental fashion starts from edge devices d_{Edge} to the optical devices $d_{Optical}$ and to the cloud data-centers. The modules that can be placed for each fog device in the path $\varepsilon d_{Edge} \cup d_{Optical}$ are identified by computing the processing requirement against the available capacity of fog devices.

A module M is placed on a fog device d_{Edge} or $d_{Optical}$ only if all other modules are already placed in the bottom-up path.

Algorithm Optical-Fog-based gaming modules placement

```
Data: M
while P ε Across all SDN paths do
    List PlacedModules;
    while Fog devices (d_Edge ∪ d_Optical) ε P do
        List ModulesToPlaced;                      /* Bottom-up traversal*/
        while module g ε GammingApp do
            if All predecessors of g are in PlacedModules then
                Add g to ModulesToPlaced;
            end
        end
    end
    while module M ε ModulesToPlace do
        if (CPU_M^req < CPU_{d_Edge}^Avail) then
            Allocate M on d_Edge ;
            if (!CPU_{d_Edge}^Avail) then
                Allocate M on d_Optical ;
            end
        else
            Allocate M on d_Optical ;
            if (!CPU_{d_Optical}^Avail)) then
                Choose M' such that (M' > M AND M'=Less-Delay-Sensitive) if (!NULL) then
                    Allocate M' on cloud data-centers;
                    Allocate M on d_Optical;
                else
                    Allocate M on cloud data-centers;
                end
            end
        end
    end
end
```

3. Performance analysis

The real-time gaming applications and CPS systems require ultralow latency, minimum energy consumption, and optimum bandwidth. The proposed Optical-Fog layer provides the desired QoE by evaluating the following parameters such as latency measure, energy consumption and bandwidth usage in contrast to the traditional cloud computing.

3.1 Latency measure

The proposed system utilizes the Optical-Fog layer that reduces the delay and improves QoE. The latency measured in the context of delay is the most concerning issue. The communication between ONU and OLT is supported by the multi-point control protocol (MPCP) which is a frame-based protocol [15]. Here, only GATE and REPORT messages are exchanged between OLT and ONU. So, in the Optical-Fog network, the delay is measured as the time between the arrival of its last bit at ONU and the arrival of its last bit at OLT. The delay $tD(f_i)$ is the computation of adding three basic components shown as:

$$tD(f_i) = \Gamma_i + t_p + T_R \tag{4}$$

Alternatively, it is the time during which the respective REPORT message reaches at the OLT completely, where Γ_i represents the one-way propagation time of ONU_i, tp represents the time between the request arriving at ONU_i and the start of the next REPORT message, and T_R represents the time duration of REPORT message [16]. Thus, the Optical-Fog layer processes more smart applications which require ultralow delay as well as efficient QoS requirements.

3.2 Energy consumption analysis

For computing energy consumption, only the edge devices of the network is a concerning issue because energy consumption by the PON channel is negligible. Thus, the total energy consumption is computed as:

$$\Delta E = E_{Edge\text{-}Fog} + E_{Optical\text{-}Fog} + E_{cloud} \tag{5}$$

- $E_{Edge\text{-}Fog} = \Sigma(E_{Edge\text{-}devices})$ represents the energy consumed by all edge devices.
- $E_{Optical\text{-}Fog} = \Sigma(E_{ONU} + E_{OLT} + E_{PON})$ represents the energy consumed by the optical elements which is negligible.
- E_{cloud} is the energy consumption of cloud data centers.

The proposed framework computes most of the computations at the Optical-Fog layer which reduces the overhead on the cloud data centers. Thus, QoE is improved by minimizing the overall energy consumption.

3.3 Bandwidth measure

Real-time applications require more bandwidth to process the extraordinarily huge volume of data. Thus, the traditional cloud system increases the overhead on communication bandwidth which results in increasing delay and poor QoE. To compute the communication bandwidth constraint, the traffic rate is assumed to be dispatched from the fog node i located at Edge-Fog layer to the server j located at cloud data center through the transmission path [17]. There is a limitation $\lambda_{i,j}^{max}$ on the bandwidth capacity of each path which is computed as:

$$0 \le \lambda_{i,j} \le \lambda_{i,j}^{max} \forall_i \in N_{fog} \text{ and } \forall_i \in M_{cloud} \quad (6)$$

Here, N_{fog} represents the set of fog devices where M_{cloud} is the set of cloud data center servers. The optimum utilization of Optical-Fog network is more effective than the traditional cloud.

4. Conclusion

The utilization of optical resources provides several benefits such as high scalability, optimum bandwidth capacity, ultralow delay, and very less energy consumption. A novel concept of FRF and ONV is used to create *OpticalFogNode* in SDN-based optical network technology. In the realization of the SDN-based optical network in fog/cloud environment, the optical network uses SDN controller efficiently to identify the shortest path (with the least congestion) to minimize delay. Also, the proposed framework effectively enhances the QoE by using the proposed module placement algorithm which enhances the QoE and makes applications more entertaining. The realization of CPS-based tasks requires optimum placement strategy which is one of the concerning issues. The proposed algorithm efficiently finds the shortest path by utilizing the concept of SDN over the optical network. The novel concept of configuring *OpticalFogNode* successfully implemented to fulfill the requirements of the CPS system. Further, the performance of the proposed system is evaluated by effectively interpreting the delay measure, bandwidth usage, and energy consumption. Finally, Optical-Fog-based deployment provides an effective platform which enhances the QoE for smart applications.

Author details

Kiran Deep Singh
Department of Computer Science and Engineering, Khalsa College of Engineering and Technology, Amritsar, Punjab, India

*Address all correspondence to: kdkirandeep@gmail.com

References

[1] Chanclou P, Cui A, Geilhardt F, Nakamura H, Nesset D. Network operator requirements for the next generation of optical access networks. IEEE Network. 2012;**26**(2):8-14

[2] Gubbi J, Buyya R, Marusic S, Palaniswami M. Internet of things (IoT): A vision, architectural elements, and future directions. Future Generation Computer Systems. 2013;**29**(7):1645-1660

[3] Bonomi F, Milito R, Zhu J, Addepalli S. Fog computing and its role in the internet of things. In: Proceedings of the First Edition of the MCC Workshop on Mobile Cloud Computing. ACM; 2012. pp. 13-16

[4] Chen S, Xu H, Liu D, Hu B, Wang H. A vision of IoT: Applications, challenges, and opportunities with China perspective. IEEE Internet of Things Journal. 2014;**1**(4):349-359

[5] Yang B, Zhang Z, Zhang K, Hu W. Integration of micro data center with optical line terminal in passive optical network. In: 2016 21st OptoElectronics and Communications Conference (OECC) Held Jointly with 2016 International Conference on Photonics in Switching (PS). IEEE; 2016. pp. 1-3

[6] Nunes BAA, Mendonca M, Nguyen X-N, Obraczka K, Turletti T. A survey of software-defined networking: Past, present, and future of programmable networks. IEEE Communication Surveys and Tutorials. 2014;**16**(3):1617-1634

[7] Sood SK, Singh KD. SNA based resource optimization in optical network using fog and cloud computing. Optical Switching and Networking. 2017

[8] Luo Y, Effenberger F, Sui M. Cloud computing provisioning over passive optical networks. In: 2012 1st IEEE International Conference on Communications in China (ICCC). IEEE; 2012. pp. 255-259

[9] Feamster N, Rexford J, Zegura E. The road to SDN: An intellectual history of programmable networks. ACM SIGCOMM Computer Communication Review. 2014;**44**(2):87-98

[10] Sood SK, Singh KD. Identification of a malicious optical edge device in the SDN-based optical fog/cloud computing network. Journal of Optical Communication. 2018. DOI: 10.1515/joc-2018-0047. [Retrieved 19 Feb, 2019]

[11] Banerjee A, Park Y, Clarke F, Song H, Yang S, Kramer G, et al. Wavelength-division-multiplexed passive optical network (WDM-PON) technologies for broadband access: A review. Journal of Optical Networking. 2005;**4**(11):737-758

[12] Khan S, Parkinson S, Qin Y. Fog computing security: A review of current applications and security solutions. Journal of Cloud Computing. 2017;**6**(1):19

[13] Sood SK, Singh KD. An optical-fog assisted EEG-based virtual reality framework for enhancing e-learning through educational games. Computer Applications in Engineering Education. 2018;**26**(5):1565-1576

[14] Yassine A, Rahimi H, Shirmohammadi S. Software defined network traffic measurement: Current trends and challenges. IEEE Instrumentation and Measurement Magazine. 2015;**18**(2):42-50

[15] Luo Y, Ansari N. Bandwidth allocation for multiservice access on EPONS. IEEE Communications Magazine. 2005;**43**(2):S16-S21

[16] Wan C-Y, Campbell AT, Krishnamurthy L. PSFQ: A reliable

transport protocol for wireless sensor networks. In: Proceedings of the 1st ACM International Workshop on Wireless Sensor Networks and Applications. ACM; 2002. pp. 1-11

[17] Deng R, Lu R, Lai C, Luan TH, Liang H. Optimal workload allocation in fog-cloud computing toward balanced delay and power consumption. IEEE Internet of Things Journal. 2016;**3**(6):1171-1181

Section 4

Advanced Relay and Antenna Systems for Smart Networks

Chapter 10

Radial Line Slot Array (RLSA) Antennas

Teddy Purnamirza

Abstract

Radial line slot array (RLSA) antenna was initially developed for satellite antenna receivers at a frequency of Ku-band. The success of this development inspired researchers to continue the study to other bands and other applications, such as Wi-Fi at 5.8 GHz. Wi-Fi applications need small antennas that lead to the diminution of RLSA antennas. Small-RLSA antennas experienced high reflection due to the small number of slots. One of the techniques that effectively eliminates the reflection was developed and named as extreme beamsquint technique. Several researches have successfully developed small-RLSA antennas by implementing this technique for Wi-Fi applications. Furthermore, for the future, it is possible to widen the researches to other frequencies and other features of RLSA antennas such as multibeam, multiband, and diminution by cutting off RLSA antennas.

Keywords: RLSA antennas, extreme beamsquint, small RLSA, RLSA for Wi-Fi, future RLSA antennas

1. Introduction

Radial line slot array (RLSA) antennas are a type of cavity or waveguide antennas. These antennas were firstly developed for satellite receivers as an option besides parabolic antennas. Unlike parabolic antennas, RLSA antennas have an advantage of having feeders at the back of the antenna, so that the feeders do not block out incoming signals. The other advantage is their flat shape so that they are more aesthetic compared to parabolic antennas. Nowadays, RLSA antennas are developed for different frequency applications such as Wi-Fi, 5th G, etc.

This chapter discusses briefly all about radial line slot array (RLSA) antennas, especially for the linearly polarized (LP)-RLSA antennas. Firstly, in Section 2, the review of RLSA antennas including the development of RLSA antennas, their applications, their development obstacles and the developed technique to overcome the obstacles are reviewed. Secondly, in Section 3, the theory of RLSA antennas is explained which includes how the antenna works and several equations to calculate antenna parameters. Thirdly, in Section 4, the mechanism of reflections in RLSA antennas, which is due to slot reflections and due to remaining power in antenna perimeter, is discussed. Fourthly, in Section 5, the theory of extreme beamsquint technique is also explained in detail. Lastly, in Section 6, the idea of future research in topic of RLSA antennas is briefly explained, including the idea of cutting off RLSA antennas to smaller size, multibeam RLSA antennas, utilizing background as

IntechOpen

radiating element and multiband RLSA antennas. It is hoped that the ideas can inspire researches for the next development of RLSA antennas.

2. Review of RLSA antenna developments

Kelly introduced the concept of RLSA antennas in the 1950s [1]. Although Kelly could produce a high-gain RLSA antenna, the structure of the antenna feeder was still complex, leading it to be costly.

In 1988, Ando et al. proposed a RLSA antenna at a frequency of 12 GHz. This antenna was designed using the technique of slot arrangements. The technique aims to produce a uniform-aperture distribution. This antenna has a double-layer cavity and exhibits a good linear polarization. Ando also proposed a beamsquint technique to improve the poor reflection coefficient in linearly polarized RLSA antennas [2, 3]. In the same year, by applying a reflection coefficient suppression and slot coupling technique, Ando successfully designed a LP-RLSA antenna for satellite applications at 12 GHz. This antenna has the efficiency of 76% and the gain of 36 dB [4–6]. Takada et al. introduced a technique to improve the reflection coefficient using a reflection cancelling slot technique. This technique successfully improved the reflection coefficient of RLSA antennas from −2 to −10 dB [7]. Endo et al. designed an optimum thickness of double-layer RLSA antennas in order to realize the mass production of thinner RLSA antennas [6].

In 1990, Ando et al. furthermore introduced a circularly polarized RLSA (CP-RLSA) antenna. This antenna utilizes a single-layer cavity instead of a double-layer cavity. This simpler cavity structure improves the complexity of RLSA fabrications and can achieve the gain of 35.4 dBi and the efficiency of 65%. Ando used two techniques to improve the antenna performance. The first is the technique of varying the slot length and slot spacing used to event out the aperture illuminations of the antenna. The second is the technique of matching spiral used to reduce the reflection of the residual power at the antenna perimeters [8, 9]. In 1991, Takashi et al. proposed the technique of varying the slot length and spacing. Utilizing this technique Takahashi proposed several high-efficiency single-layer RLSA antennas with the diameter of 25–60 cm. These antennas can achieve efficiencies of between 70 and 84% [10]. Furthermore, Takashi et al. produced and marketed a-78% efficiency, 32.6 dB gain and single-layer RLSA [11–13].

Australian researchers started to investigate RLSA in 1995. They reported several investigations to design LP-RLSA antennas for satellite receivers. These investigations used the combination of the theoretical and experimental approach. The availability of low-cost materials (polypropylene) and low-cost fabrication also become a consideration in these researches. In 1997, Davis reported a 60 cm diameter LP-RLSA prototype designed using the reflection cancelling slot technique. This technique can overcome the inherent poor reflection coefficient of LP-RLSA antennas [4]. Davis and Bialkowski also successfully tested a RLSA antenna designed utilizing the reflection cancelling slot technique and a beamsquint value of 20° [14, 15]. Furthermore, Davis and Bialkowski reported an investigation of LP-RLSA antennas utilizing the beamsquint technique for several squint angles. This technique successfully improved the reflection coefficient under −25 dB [16]. Davis integrated the report of [2, 4, 7, 11] to form a beam synthesis algorithm used to calculate the design parameter of LP-RLSA antennas [17].

Due to the successful development of RLSA antennas for satellite applications, researchers tried to bring RLSA antennas into small antenna application for Wi-Fi devices. However, the design of small-RLSA antennas was not easy since small-size RLSA antennas normally performed high reflection coefficient [18–20]. Hirokawa et al. used a technique for matching slot pair in order to reduce the remaining power at the antenna perimeter of small-aperture RLSA, so that this technique can minimize the reflection coefficient [21, 22]. Akiyama et al. also used the same technique for matching slot pair [23, 24]. However, the technique for matching slot pair is only used to radiate the remaining power at the antenna perimeter and does not contribute to the antenna gain. Reference [25] introduced the use of long slots in order to increase the ability of slots to radiate power, so that it can reduce the remaining power at the perimeter of small-aperture RLSA antennas, thus reducing the reflection coefficient. However, although this method can reduce the reflection coefficient, this method also can decrease the antenna gain. This is because that the long slots cannot radiate a focus power.

In 2002, Malaysian and Australian researchers started to investigate the application of RLSA antennas for wireless LANs. Tharek and Ayu successfully fabricated a low-profile RLSA antenna at a frequency of 5.5 GHz with a broad radiation pattern of 60° used for indoor wireless LANs [26]. Bialkowski and Zagriatski investigated the design of RLSA antennas for wireless LANs and successfully fabricated a dual-band 2.4/5.2 GHz antenna [27, 28]. Furthermore, Imran et al. reported the design and test of RLSA antennas for outdoor point-to-point applications at the frequency of 5.8 GHz [29–31]. However, this design utilized a beamsquint technique that is similar with the technique used to design RLSA antennas for satellite applications. Hence, the diameter of this antenna is still considered large with a diameter of 650 mm, so that it is not applicable for small Wi-Fi devices. Islam reported the utilization of low-cost FR4 materials to fabricate RLSA antennas at the frequency of 5.8 GHz for wireless LANs. This invention is quite innovative since FR4 materials are a low-cost material and easy to be fabricated [32, 33]. However, there are some drawbacks in designing this antenna, such as a design of overlap slots, a loss cavity due to the use of several FR4 boards and the use of material loss of FR4. These all lead to low gain (only 8 dB) and low bandwidth (75 MGhz).

Purnamirza et al., in 2012, introduced a technique called extreme beamsquint technique in order to overcome the problem of high reflection in small-RLSA antennas [34]. This technique uses the beamsquint values higher than 60°. The theory of how the high values of beamsquint can significantly minimize the reflection coefficient is explained. Purnamirza also developed RLSA antennas that mimic the specification of other types of antenna that is available in markets [35–38].

3. Basic theory of RLSA antennas

This section discusses the theory of RLSA antennas including the structure, the theory of how RLSA antennas work as well as several formulas to design RLSA antennas.

3.1 Structure of RLSA antennas

Figure 1 shows the illustration of the structure of a RLSA antenna. The figure shows the structure of RLSA antennas consisting of a radiating element, a cavity, a background and a feeder. The radiating element usually is a circular plate made of metals, such as aluminium, copper or brass. The radiating element consists of many slot pairs. One slot pair acts as one antenna element so that all the slot pairs form an array antenna. The background is a metal plate just like the radiating element, but the background does not have slots. The cavity is a dielectric material that has the form of a tube. Together with the radiating element and the background, the cavity operates as a circular waveguide that guides the signal from the feeder to propagate in radial direction. The feeder is a part of RLSA antennas used to feed signals from a transmission line into the antenna.

3.2 How RLSA antennas work

Figure 2 shows the wave propagation mechanism including TEM cavity mode and TEM coaxial mode. The feeder placed in the centre of the antenna cavity feeds the electromagnetic power (indicated by the arrows). The feeder is an ordinary SMA feeder, which is modified by adding a head disc. The head disc has a function to convert the electromagnetic power from a TEM coaxial mode into a TEM cavity mode (a radial mode), so that the electromagnetic power fed by the feeder will propagate in a TEM mode and in a radial direction within the antenna cavity.

When the power passes the slot pair, some amount of the power escapes through the slot pair and radiates as illustrated in **Figure 3**. Hence, the slot pair can be considered as one antenna element. Since there are many slot pairs (thousands in normal-size RLSA antennas), all the slot pairs will form an array antenna. Therefore, this is the reason why 'array' word is included in the name of RLSA.

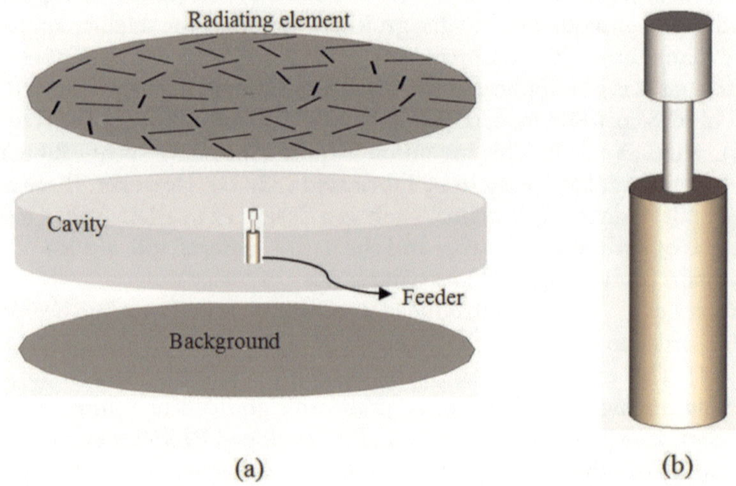

Figure 1.
(a) The component of RLSA antennas. (b) The magnified view of the feeder [39].

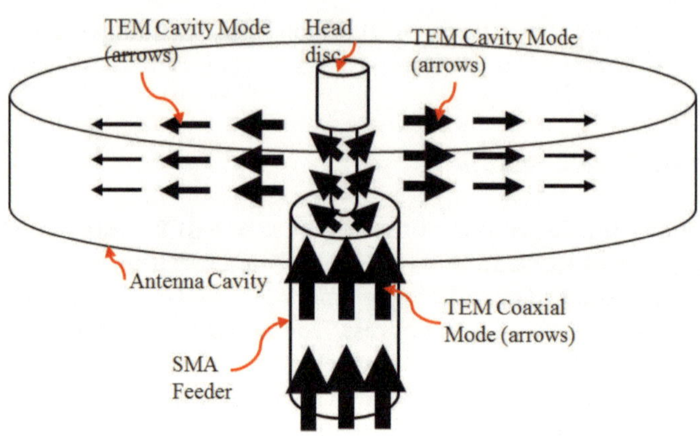

Figure 2.
Illustration of the TEM cavity mode and the TEM coaxial mode [39].

3.3 Polarizations

A slot pair, which represents a signal source in RLSA antennas, is located in the top surface of the radiating element of a RLSA antenna. A linear polarization in the RLSA antenna can be produced by combining two signals from the slot pair. **Figure 4a** shows the illustration of the slot pair. The signal from Slot 1 and the signal from Slot 2 have a phase difference of 180° or phi radians since Slot 1 and Slot 2 have the distance of half wavelength ($0.5\lambda_g$) to each other. Since the orientation of Slot 1 and Slot 2 is perpendicular to each other, the signals from Slot 1 (at y axis) and Slot 2 (at x axis) are also perpendicular to each other, as shown in **Figure 4b**.

Figure 4b shows that when Signal 1 is increasing in positive values, Signal 2 is decreasing in negative values. Since their position is perpendicular to each other, the resulting wave becomes a line in Quadrant II. When Signal 1 is decreasing towards zero and Signal 2 is increasing towards zero, the resulting signal will be a line in Quadrant II but with a shorter length compared to the line in the previous case. When Signal 1 is decreasing in negative values and Signal 2 is increasing in positive values, then the resulting signal will be a line in Quadrant IV. When Signal 1 is increasing towards zero and Signal 2 is decreasing towards zero, then the resulting signal will be a line in Quadrant IV but with the shorter length compared to the line in the previous case. Now, we can understand that the resulting signal of Signal 1 and Signal 2 results in a signal that looks like a straight line where the

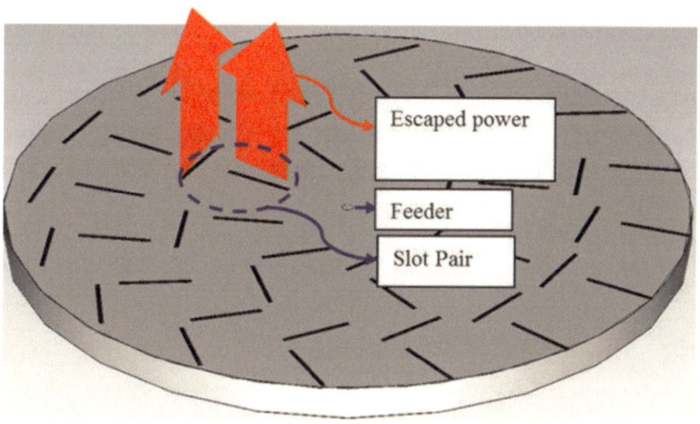

Figure 3.
Illustration of the power escaping from the slot pairs [39].

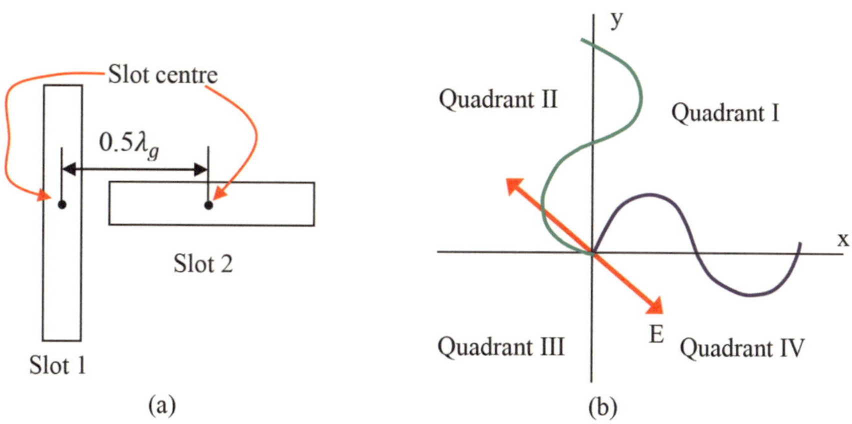

Figure 4.
Polarization establishment in a linearly polarized RLSA [39]. (a) slot pair position (b) signal of each slot.

length changes as a function of time; this is the reason why its name is 'linear polarizations'.

3.4 Orientation of the slot in RLSA antennas

Figure 5 shows the position of the slots (indicated by 'A' and 'B') and the squint of the inclination angles of the slots (indicated by 'θ1 and θ2). The slot pair must be located in the correct position on the radiating surface of RLSA antennas. The slot pair must be located in different and unique positions in order to prevent overlapping between them.

Equations (1) and (2) express the squint of the slots obtained by the beamsquint technique [4, 14–17, 39–43]:

$$\theta_1 = \frac{\pi}{4} + \frac{1}{2}\left\{\arctan\left(\frac{\cos(\theta_T)}{\tan(\phi_T)}\right) - (\phi - \phi_T)\right\} \tag{1}$$

$$\theta_2 = \frac{3\pi}{4} + \frac{1}{2}\left\{\arctan\left(\frac{\cos(\theta_T)}{\tan(\phi_T)}\right) - (\phi - \phi_T)\right\} \tag{2}$$

where θ_1 is the inclination angle of Slot 1; θ_2 is the inclination angle of Slot 2; θ_T is the beamsquint angle in elevated direction; ϕ is the azimuth angle of Slot 1 and Slot 2 position; and ϕ_T is the beamsquint angle in azimuth direction.

3.5 Arrangement of slot pairs

Figure 6 shows the geometrical arrangement of a slot pair or also called a unit radiator. The arrangement of the unit radiator in the radiating surface of RLSA antennas must be carefully calculated and drawn since a little deviation of the unit radiator position will rapidly decrease the performance of RLSA antennas.

Based on **Figure 6**, the distance of a particular unit radiator from the centre point of RLSA antennas is expressed in Eq. (3) [4, 14–17, 39–43]:

$$\rho_p = \frac{n\lambda_g}{1 - \xi sin\theta_T \cos(\phi - \phi_T)} \tag{3}$$

Where $\xi = \frac{1}{\sqrt{\varepsilon_{r.}}}$

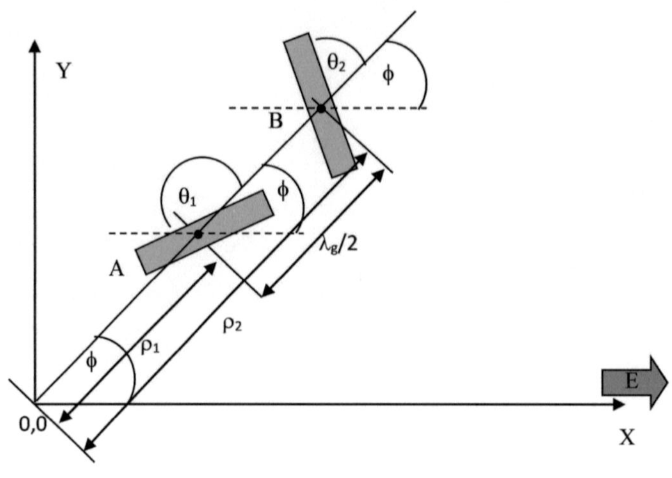

Figure 5.
Slot pair geometry [39].

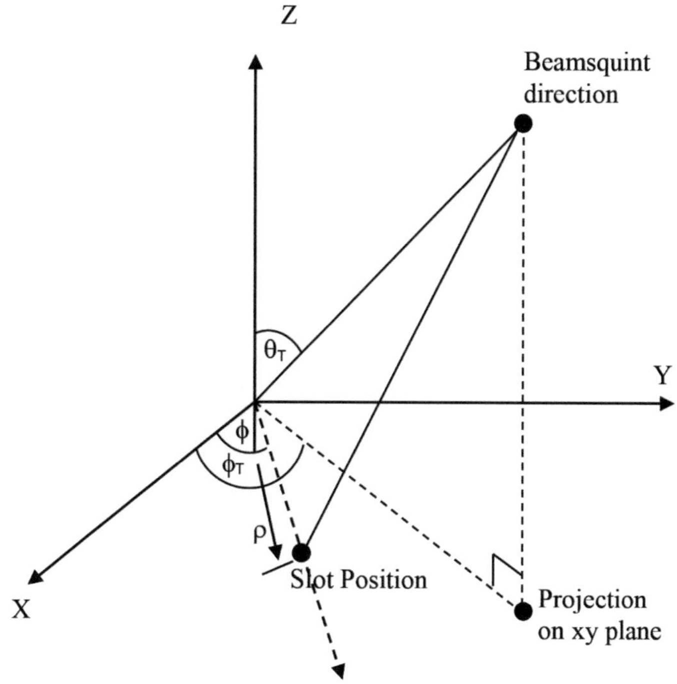

Figure 6.
Geometrical arrangement of a unit radiator [39].

Equation (4) expresses the distance between two adjacent unit radiators located in two different rings (the distance in the radial direction) [4, 14–17, 39–43]:

$$S_\rho = \frac{\lambda_g}{1 - \xi sin\theta_T \cos\left(\phi - \phi_T\right)} \tag{4}$$

Equation (5) expresses the distance between two adjacent unit radiators in a same ring (the distance in the azimuth direction) [4, 14–17, 39–43]:

$$S_\phi = \frac{2\pi\lambda_g}{\sqrt{1 - \xi^2 \sin {\theta_T}^2}} \frac{q}{p} \tag{5}$$

where λ_g is the length of the wavelength inside the cavity of RLSA antennas; ε_r is the relative permittivity of the cavity of RLSA antennas.

n is the ring numbers (1,2,3, etc.); q is the integer numbers (1, 2, 3, etc.) that express the distance of the innermost ring from the centre of RLSA antennas; and p is the number of unit radiators in the innermost ring.

The parameters of S_ρ, S_ϕ, ρ_ρ,ρ_1, and ρ_2 are shown in **Figure** 7. Since the distance from the centre of the unit radiator to Slot 1 or Slot 2 is '$\lambda_g/4$', Eqs. (5)–(7) express the distance of slots from the centre of antennas [4, 14–17, 39–43]:

$$\rho_{\rho 1} = \frac{(n - 1 + q - 0.25)\lambda_g}{1 - \xi sin\theta_T \cos\left(\phi - \phi_T\right)} \tag{6}$$

$$\rho_{\rho 2} = \frac{(n - 1 + q + 0.25)\lambda_g}{1 - \xi sin\theta_T \cos\left(\phi - \phi_T\right)} \tag{7}$$

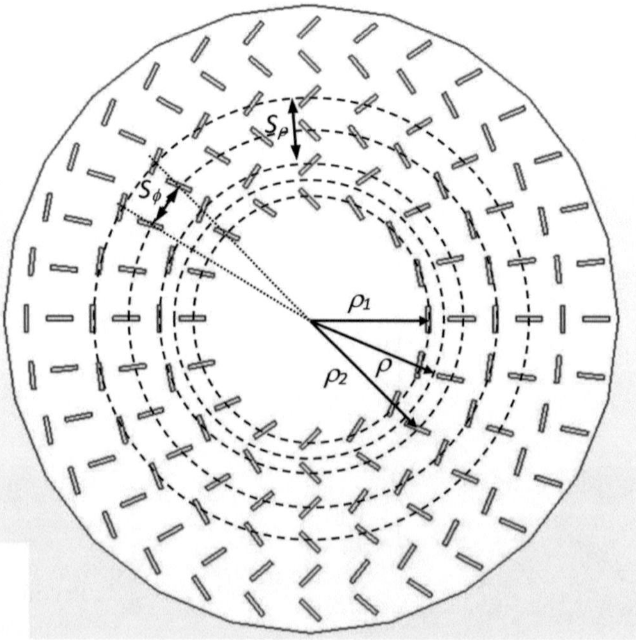

Figure 7.
Definition of some slot parameters [39].

3.6 Length of slots

The length of the slots on the radiating surface of RLSA antennas must be varied in order to achieve a uniform aperture illumination. The farther a slot from the centre of the antenna, the longer the length of the slot will be. The length of the slot is the function of ρ that is the distance of the slot from the centre of the antennas, as expressed by Eq. (8) [42]:

$$L_{rad} = \left(4.9876 \times 10^{-3}\rho\right)\frac{12.5\,x10^9}{f_0} \tag{8}$$

The formula in Eq. (8) is an approximate formula. To get an accurate formula, we need to do some measurements and experiments.

4. Reflection in small-RLSA antennas

4.1 Signal reflection due to remaining power

The power (P) comes from the feeder, which is located at the centre of the antenna, and flows towards the antenna perimeter, as illustrated in **Figure 8b**. When the power passes the slots, some amount of the power radiates through the slots. The power inside the cavity will decrease every time the power passes the slots and will continue to decrease until the power reaches the antenna perimeter. Equation (9) expresses the remaining power (P_R) at the antenna perimeter [42]:

$$(P_R) = P\,(1-\alpha)^n \tag{9}$$

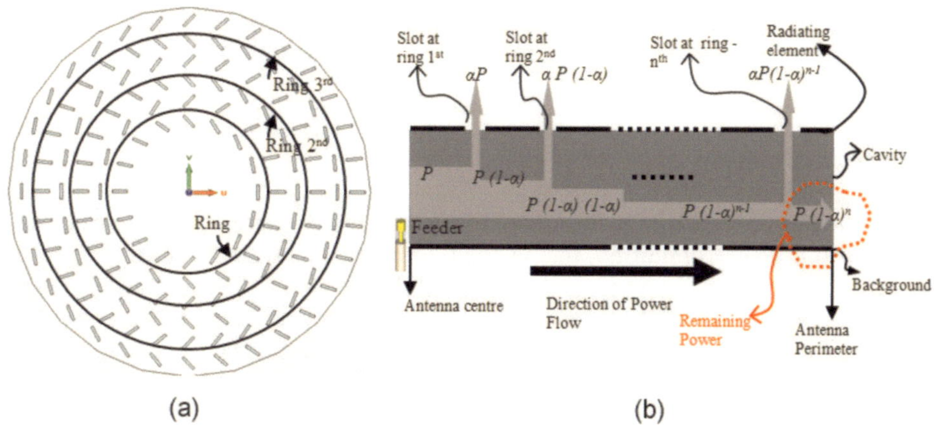

Figure 8.
(a) Top view of RLSA. (b) Cut view of a RLSA antenna and the power flow mechanism inside the RLSA cavity [39].

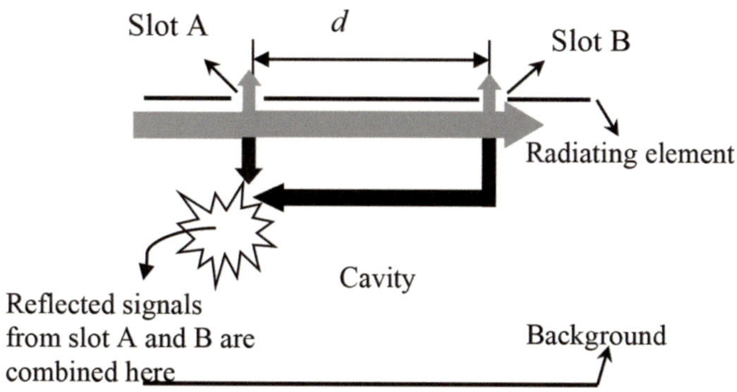

Figure 9.
Illustration of the reflected signals from the slot [39].

Equation (9) shows that the amount of the remaining power depends on the number of rings (n), which is also proportional to the number of slots. For small-RLSA antennas, which have a small number of slots, the amount of the remaining power at the antenna perimeter will be high. Part of this remaining power will be reflected back to the feeder and result in a high signal reflection, thus increasing the reflection coefficient. For normal-size RLSA antennas, which have thousands of slots, the remaining power at the antenna perimeter is very small so that its effect to the signal reflection is neglected.

4.2 Signal reflection due to the reflected signal from slots

Figure 9 shows the front cut view of a RLSA antenna and the signal flow within the cavity of the RLSA antenna. The grey arrows represent the signals that flow from the centre of the RLSA antenna to the antenna perimeter, and the black arrows represent the reflected signal from the slots. **Figure 9** shows that since the distance between the slots (d) is $\lambda_g/2$, the signal from slot 'A' will travel for $\lambda_g/2$ to reach 'B'. At 'B', some of the signal will be reflected back and travel for another $\lambda_g/2$ to reach 'A'. Therefore, the reflected signal from slot 'A' and slot 'B' will have a different phase of $\lambda_g/2 + \lambda_g/2 = \lambda_g$ or 360° (or can be said there is no phase difference), so that they will strengthen each other and result in a high signal reflection [42].

5. Extreme beamsquint technique

The ability of beamsquint technique in minimizing the reflected signal from the slots depends on one condition, that is, the number of ring must be sufficient. As an example, **Figure 10a** and **b** shows the reflected signals of a three-ring RLSA antenna and the reflected signals of a two-ring RLSA antenna, respectively. Since every ring consists of two slots, hence, there are six reflected signals for the three-ring RLSA antenna and four reflected signals for the two-ring RLSA antenna. It is assumed that the amplitude of all reflected signals is the same in order to simplify the analysis. From **Figure 10a**, it can be observed that all the graph space is covered by the reflected signals; hence the combination of all reflected signals will cancel out each other, and the minimum signal reflection is obtained. In contrast, from **Figure 10b**, it can be seen that not all graph space (the area pointed by 'A') is covered by the reflected signals; hence the combined signal will be greater than the combined signal in **Figure 10a**.

From the example in the previous paragraph, it can be concluded that a smaller number of ring will decrease the ability of beamsquint technique in cancelling the reflected signal. Therefore, this is the reason why the reflection coefficient of small-RLSA antennas, which have few numbers of rings (less than 2), is high and why the normal beamsquint technique fails to minimize the reflection coefficient of small-RLSA antennas. The next section will explain how the proposed extreme beamsquint technique can reduce the high reflection coefficient of small-RLSA antennas by increasing the number of ring.

The position of the ring in radial direction (S_ρ) can be expressed by Eq. (10) [21]:

$$S_\rho = \frac{r\lambda_g}{1 - \xi \sin\theta_T \cos(\phi - \phi_T)} \tag{10}$$

θ_T is the beamsquint angle, ϕ is the position of slots in azimuth, ϕ_T is the azimuth angle of beamsquint and r is the ring number. **Figure 11a** illustrates the definition of all this parameters.

Based on Eq. (10), by utilizing $r = 1$, $\phi_T = 0$ and $\phi = 0–360°$, the rings for beamsquint angle of 10°, 30° and 60° are plotted as shown in **Figure 11b**.

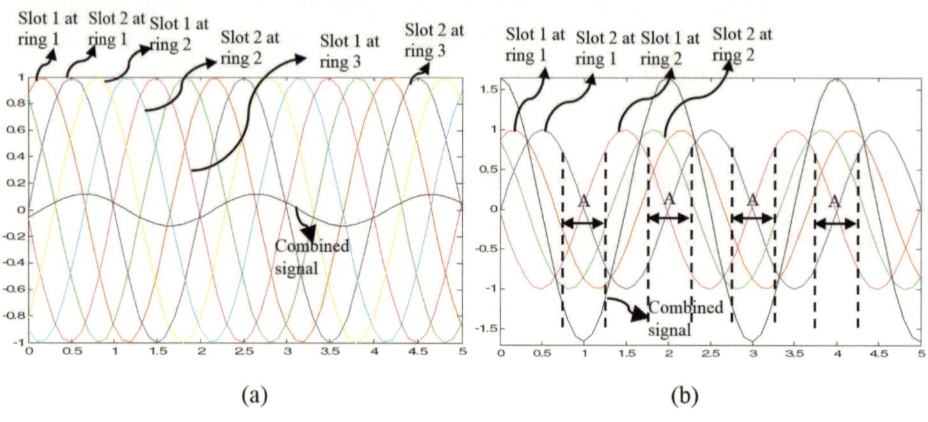

Figure 10.
(a) Reflected signal of a three-ring RLSA antenna. (b) Reflected signal of a two-ring RLSA antenna [39].

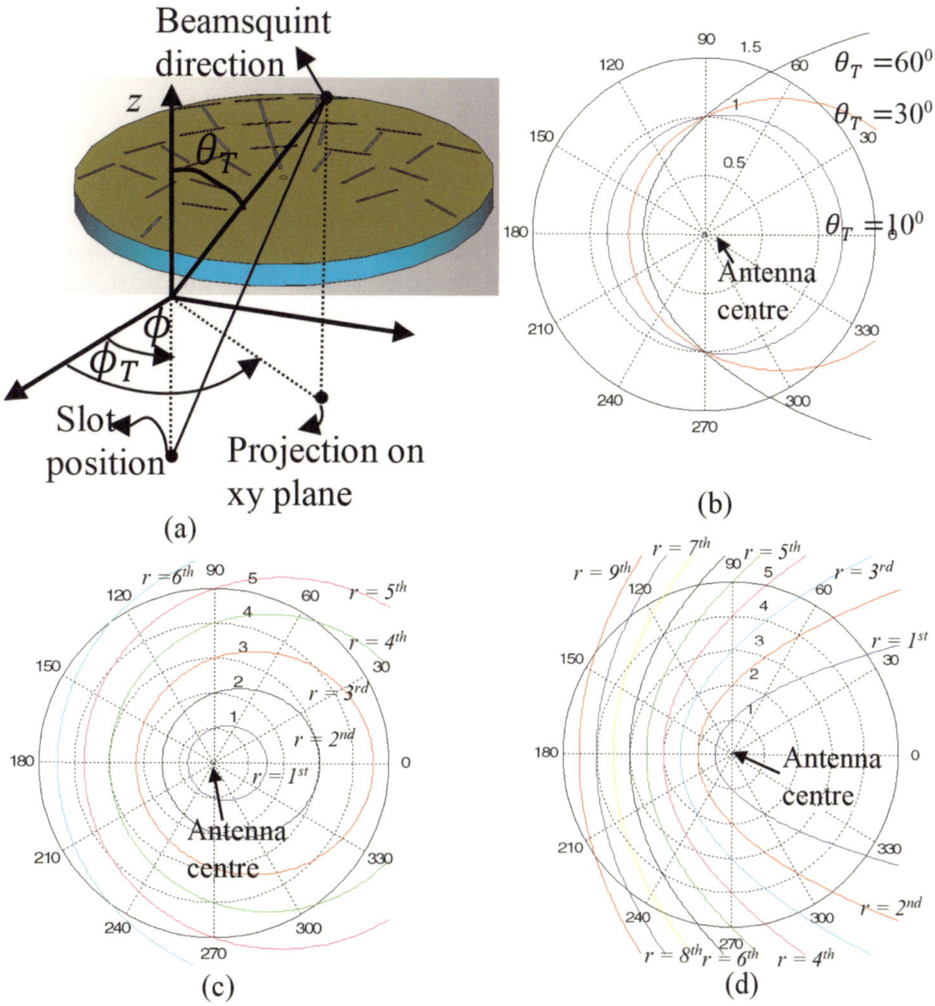

Figure 11.
(a) Illustration of some parameters of ring position. (b) Plot of the ring for beamsquint of 20, 30 and 60°. (c) Plot of various ring numbers for beamsquint angle of 20°. (d) Plot of various ring numbers for beamsquint angle of 80° [39].

It illustrates that the beamsquint technique performs the ring in the shape of ellipse rather than in the shape of circular. From **Figure 11b**, it can be observed that the position of the ring at the left-hand side will move closer to the centre of the antenna as the beamsquint increases. In contrast, the position of the ring at the right-hand side will move farther from the centre of the antenna as the beamsquint increases.

Still based on Eq. 3, by utilizing $\phi_T = 0$ and ϕ = 0 to 360°, the rings are plotted for various ring numbers both for the beamsquint angle of 20° and 800 as shown in **Figure 11c** and **d**, respectively. From these figures, it can be observed that at the left-hand side, the distance between rings for beamsquint angle of 80° is shorter than the distance between the rings for beamsquint angle of 20°. Due to the shorter distance between rings, the beamsquint angle of 80° has more rings (nine rings) that can be plotted in the antenna area compared to the beamsquint angle of 20° (six rings). Based on the previous examples and explanations, it can be concluded that the higher beamsquint angle can yield more rings. This fact is very useful to include additional rings for the small-RLSA antenna, which originally has a low number of rings (less than 2). The extra number of rings will have more ability to minimize the reflection coefficient of small-RLSA antenna. The use of extra high beamsquint angle underlies the naming of extreme beamsquint technique.

6. Future development of RLSA antennas

Space to be explored for RLSA antennas has remained wide, since only few researchers study this topic. This is due to the drawing slots of RLSA antennas that are difficult without using computer programmes. This is unlike microstrip antennas which are easy to draw since their shape is simple. Moreover, it is more difficult to fabricate RLSA antennas compared to microstrip antennas since there is no row material that is ready to be cut or to be formed. This is unlike microstrip, which can use many types of boards such as FR4. Therefore, due to these reasons, less researchers are interested in studying RLSA antennas than microstrip antennas. Below, several research ideas in the field of RLSA are presented to be explored deeper, especially for doctoral dissertation and master thesis.

6.1 Cutting antennas

6.1.1 Concentrated slot area

The use of extra high beamsquint results in concentrated slots in a certain area of radiating elements, and leaving other areas vacant from slots, as shown in **Figure 12a**.

Our hypothesis is that since the vacant area is not useful, then it can be cut off, thus resulting in a smaller antenna, as shown in **Figure 12b**. We have studied that definitely there will be an effect of the cut, which is a leakage power along the cutting line, shown in **Figure 12c**. This leakage power reduces the antenna gain. However, the antenna gain will not be affected significantly since the power density within the cut antenna will increase, thus also will increase the gain, then countering the decrease gain due to the leakage power.

6.2 The use of background as radiating element

Theoretically, only radiating elements that are used as place for slots, as shown in **Figure 13a**. The background always functions as conventional background to

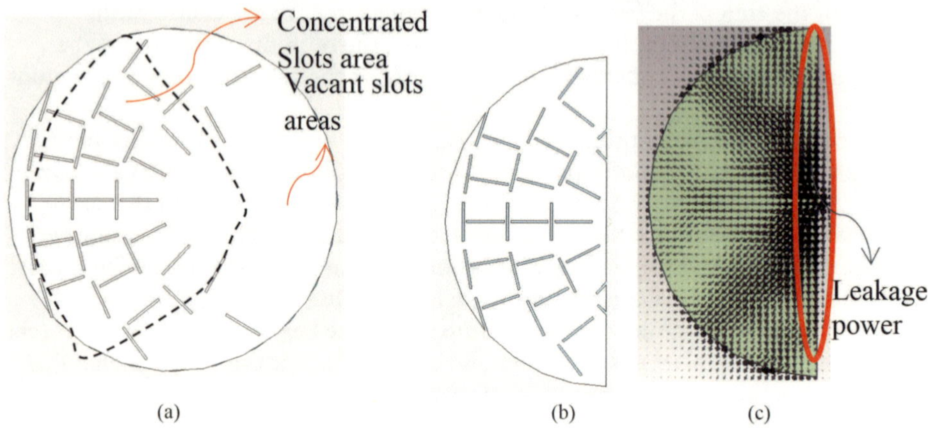

Figure 12.
(a) Concentrated slots. (b) Cut antennas. (c) Leakage power.

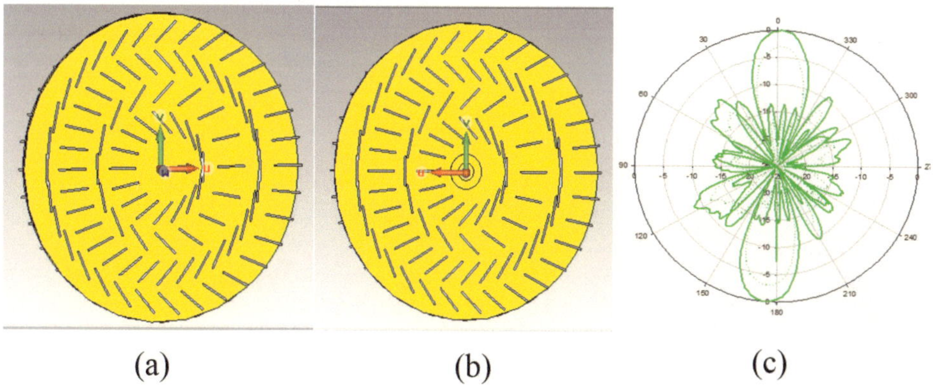

Figure 13.
(a) Radiating element. (b) Background. (c) Radiation pattern.

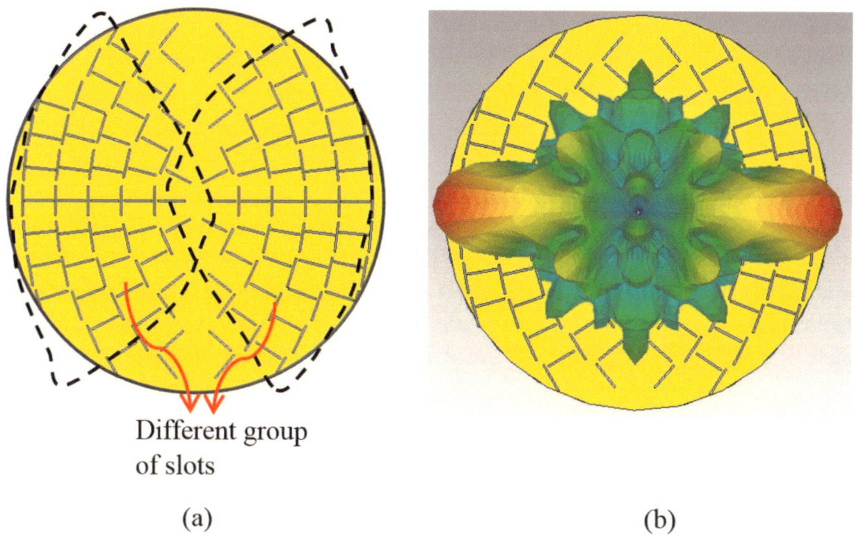

Figure 14.
(a) Design of dual-beam antenna. (b) Beam result.

radiating elements. Our hypothesis is that the background can be used to draw slots on it, as shown in **Figure 13b**. This will result in a dual-beam antenna as shown in **Figure 13c**. Of course, the gain will decrease by 3 dB compared to the originally single-beam antenna, since the power is divided into two beams.

6.3 Multibeam antennas

Multibeam antennas can be produced by designing slots using different beamsquint values. The slots are grouped and placed based on its beamsquint values. **Figure 14a** shows the slot design of two beams, and **Figure 14b** shows its beams. Things to note are that there will be a slot coupling between adjacent slot groups, which will lessen the sharpness of beams or the gains. Therefore, it is needed to put the slot group not too close together such that we get satisfied gains.

6.4 Multiband antennas

Multiband antennas can be produced by designing slots using different frequencies. **Figure 15** shows the slots that are used for two frequencies. It is noted that we

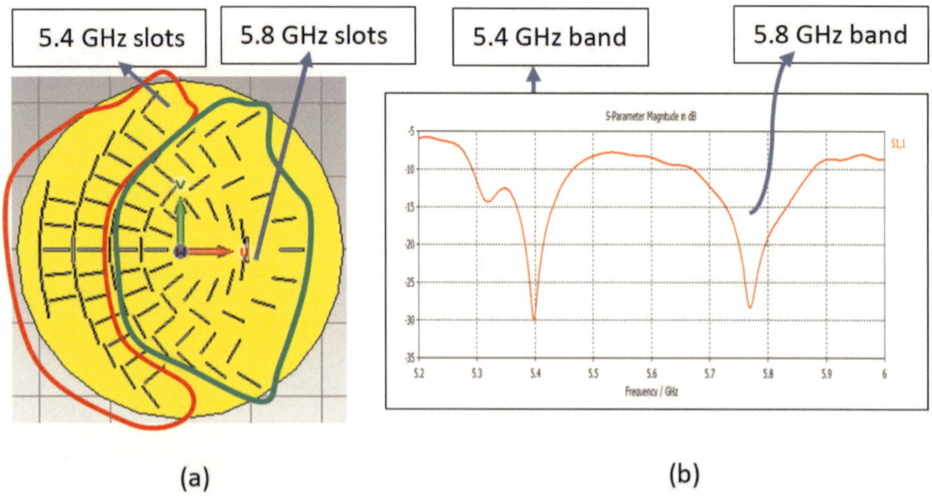

Figure 15.
(a) Design of dual-band antenna. (b) Band result.

cannot differentiate between the 5.4 and 5.8 GHz slots since they are plotted mixed. This will not effect to the gain since all slots are designed using a same beamsquint value. Research on this topic can be as follows: firstly, a study on how to correctly place different frequency slots, so that the correct placement will not increase antenna reflections. Secondly, a study of how far two different bands or more can be separated since a different band needs a different feeder structure. However in multiband antennas, we use a same feeder structure so that it will effect to the increase of reflection, such that different bands cannot be separated too far.

7. Conclusions

RLSA antennas have been developed since the 1980s until nowadays. They were initially developed for satellite receiver antennas. Due to their advantages, such as high gains and flat shapes, these antennas have been also developed for small device applications at smaller frequencies, such as Wi-Fi, 5th G, 4th G, etc. However developing RLSA antennas for small device applications had been facing the problem of high signal reflections, due to the limited number of slots. Several techniques had been proposed to overcome the problem. Among the techniques, the extreme beamsquint technique is the most effective technique in reducing signal reflections. By resolving the problem of signal reflections, the research areas for RLSA antennas become wide open, especially for multibeam RLSA antennas, multiband RLSA antennas and size minimization by cutting RLSA antennas.

Author details

Teddy Purnamirza
Universitas Islam Negeri Sultan Syarif Kasim, Pekanbaru Riau, Indonesia

*Address all correspondence to: tptambusai@uin-suska.ac.id

References

[1] Kelly KC. Recent annular slot array experiments. In: IRE National Convention Record; Vol. 5; 1957;**1**. pp. 144-151

[2] Ando M et al. Linearly polarized radial line slot antenna. IEEE Transactions on Antennas and Propagation. 1988;**36**:1675-1680

[3] Ando M et al. Linearly-polarized radial line slot antenna. IEEE Antennas and Propagation Society. 1988;**1**:836-839. AP-S International Symposium (Digest)

[4] Davis PW. Experimental investigations into a linearly polarized radial slot antenna for DBS TV in Australia. IEEE Transactions on Antennas and Propagation. 1997;**45**: 1123-1129

[5] Ando M et al. Radial line slot antenna for DBS reception. In: Conference Proceedings—European Microwave Conference; 1988. pp. 306-311

[6] Endo K et al. Waveguide design of a radial line slot antenna. Electronics and Communications in Japan. Part I: Communications. 1990;**73**:109-115. (English translation of Denshi Tsushin Gakkai Ronbunshi)

[7] Takada J-i et al. A reflection cancelling slot set in a linearly polarized radial line slot antenna. IEEE Transactions on Antennas and Propagation. 1992;**40**:433-438

[8] Ando M et al. Single-layered radial line slot antenna for DBS reception. In: Conference Proceedings—European Microwave Conference; 1990. pp.1541-1546

[9] Ando M et al. A slot design for uniform aperture field distribution in single-layered radial line slot antennas. IEEE Antennas and Propagation Society. 1990;**1**:930-933. AP-S International Symposium (Digest)

[10] Takahashi M et al. High efficiency flat array antennas for DBS reception. In: Conference Proceedings—European Microwave Conference; 1991. pp. 629-634

[11] Takahashi M et al. A slot design for uniform aperture field distribution in single-layered radial line slot antennas. IEEE Transactions on Antennas and Propagation. 1991;**39**:954-959

[12] Takahashi M et al. Characteristics of small-aperture, single-layered, radial-line slot antennas. IEEE Proceedings H: Microwaves, Antennas and Propagation. 1992;**139**:79-83

[13] Takahashi M et al. Small aperture single-layered radial line slot antenna for DBS reception. In: IEEE Conference Publication; 1991. pp. 738-741

[14] Davis PW, Bialkowski ME. Comparing beam squinting and reflection cancelling slot methods for return loss improvement in RLSA antennas. IEEE Antennas and Propagation Society. 1997;**3**: 1938-1941. AP-S International Symposium (Digest)

[15] Davis PW, Bialkowski ME. Linearly polarized radial-line slot-array antennas with improved return-loss performance. IEEE Antennas and Propagation Magazine. 1999;**41**:52-61

[16] Davis PW, Bialkowski ME. Performance of a linearly polarized RLSA antenna for different beam squint angles. In: Asia-Pacific Microwave Conference Proceedings (APMC); 1997. pp. 653-656

[17] Davis PW, Bialkowski ME. Beam synthesis in linearly polarized radial line slot array antennas. IEEE Antennas and

Propagation Society. 2000;**1**:94-97. AP-S International Symposium (Digest)

[18] Tong KF et al. Broadband characteristic of an offset aperture-coupled dual circular disk microstrip antenna. Electronics Letters. 1994;**30**: 1728-1729

[19] Mowlér M, Lindmark B. A matching slot pair for a circularly-polarized slotted waveguide array. Proceeding in 2006 First European Conference on Antennas and Propagation; 2006.1

[20] Haridas N et al. Reconfigurable MEMS antennas. In: Proceedings of the NASA/ESA Conference on Adaptive Hardware and System; 2008. pp. 147-154

[21] Hirokawa J et al. Matching slot pair for a circularly-polarized slotted waveguide array. IEEE Proceedings H: Microwaves, Antennas and Propagation. 1990;**137**:367-371

[22] Hirokawa J et al. A matching slot 21 pair for a circularly-polarized slotted 22 waveguide array. Proceeding in International Symposium on Antennas and Propagation Society, Merging Technologies for the 90's. 1990;**1**. pp. 926-929

[23] Akiyama A et al. Numerical optimization of slot parameters for a concentric array radial line slot antenna. IEEE Proceedings: Microwaves, Antennas and Propagation. 1998;**145**: 141-145

[24] Akiyama A et al. Design of radial line slot antennas for millimeter wave wireless LAN. IEEE Antennas and Propagation Society. 1997;**4**:2516-2519. AP-S International Symposium (Digest)

[25] Zagriatski S, Bialkowski ME. Circularly polarized radial line slot array antenna for wireless LAN access point. In: 15th International Conference on Microwaves, Radar and Wireless Communications (MIKON—2004); 2004. pp. 649-652

[26] Tharek AR, Ayu IKF. Theoretical investigations of linearly polarized radial line slot array (RLSA) antenna for wireless LAN indoor application at 5.5 GHz. In: Proceedings of the Mediterranean Electrotechnical Conference—MELECON; 2002. pp. 364-367

[27] Bialkowski KS, Zagriatski S. Investigations into a dual band 2.4/ 5.2 GHz antenna for WLAN applications. In: 15th International Conference on Microwaves, Radar and Wireless Communications (MIKON-2004); 2004. pp. 660-663

[28] Bialkowski KS, Zagriatski S. A dual band 2.4/5.2 GHz antenna including a radial line slot array and a patch. IEEE Antennas and Propagation Society. 2004;**3**:3095-3098. AP-S International Symposium (Digest)

[29] Imran MI, Tharek AR. Radial line slot antenna development for outdoor point to point application at 5.8 GHz band. In: 2004 RF and Microwave Conference, RFM 2004—Proceedings; 2004. pp. 103-105

[30] Imran MI et al. An optimization of beam squinted radial line slot array antenna design at 5.8 GHz. In: 2008 IEEE International RF and Microwave Conference (RFM 2008); Sabah; 2008. pp. 139-142

[31] Imran MI et al. Beam squinted radial line slot array antenna (RLSA) design for point-to-point WLAN application. In: 2007 Asia-Pacific Conference on Applied Electromagnetics Proceedings (APACE-2007); 2007

[32] Islam MRU, Rahman TA. Novel and simple design of multi layer radial line slot array (RLSA) antenna using FR-4 Substrate. In: 2008 Asia-Pacific Symposium on Electromagnetic

Compatibility and 19th International Zurich Symposium on Electromagnetic Compatibility (APEMC 2008); 2008. pp. 843-846

[33] Islam MRU et al. Simple integrated system for wireless backhaul networks. In: Proceedings of the International Conference on Computer and Communication Engineering 2008 (ICCCE08): Global Links for Human Development; 2008. pp. 341-345

[34] Purnamirza T, Tharek AR, Jamaluddin MH. The extreme beamsquint technique to minimize the reflection coefficient of very small aperture-radial line slot array antenna. Journal on Electromagnetic Wave and Application. 2012;**26**: 2267-2276

[35] Purnamirza T, Kristanto D, Ibrahim IM. A design of compact radial line slot array (RLSA) antennas for Wi-Fi market needs. Progress in Electromagnetics Research Letter. 2016; **64**:21-28

[36] Purnamirza T, Ibrahim IM, Prowadi P, Amillia F. Small radial line slot array (RLSA) antennas for Wi-Fi 5.8 GHz devices. International Journal on Communications, Antenna and Propagation (IRECAP). 2017:397-402

[37] Purnamirza T, Hasbi S, Ibrahim IM, Mulyono, Amillia F, Rahmi D. A radial line slot array (RLSA) antenna with the specifications of 16 dBi outdoor patch antenna. TELKOMNIKA (Telecommunication Computing Electronics and Control). 2018;**7**:46-52

[38] Purnamirza T, Budikesuma P, Ibrahim IM, Rahmi D, Susanti R. A small-RLSA antenna utilizing the specification of back fires 17 dBi LAN antennas. Telkomnika (Telecommunication Computing Electronics and Control). 2018;**16**:16

[39] Purnamirza T. Very small aperture radial line slot array beam steering antennas [thesis]. Johor: Universiti Teknologi Malaysia; 2013

[40] Bialkowski ME, Davis PW. Predicting the radiation pattern of a radial line slot array antenna. In: Asia-Pacific Microwave Conference Proceedings (APMC); 1999. pp. 162-165

[41] Bialkowski ME, Davis PW. Design and development of a radial line slot array antenna of arbitrary polarization. In: Asia-Pacific Microwave Conference Proceedings (APMC); 2000. p. 2

[42] Bialkowski ME, Davis PW. Linearly polarized radial-line slot-array antenna with a broadened beam. Microwave and Optical Technology Letters. 2000;**27**: 98-101

[43] Davis PW. A linearly polarised radial line slot array antenna for direct broadcast satellite services [thesis]. Queensland: University of Queensland; 2000

Chapter 11

New High-Speed Directional Relay Based on Wireless Sensor Network for Smart Grid Protection

Ali Hadi Abdulwahid

Abstract

The production of energy from water represents large amounts of clean and renewable energy. However, only 30% of this energy has been developed so far. Hydropower, particularly hydropower plants, is not only environmentally friendly but also economical, and operates more efficiently than any other renewable energy system. Hydropower plants are largely automated and have relatively low operating costs. The main components of the power system must be continuously monitored and protected to maintain the quality and reliability of the power source. This task is provided by the data collection, monitoring and protection system. Turbines must be protected not only by short circuits but also by abnormal conditions. The proposed protection has been designed to avoid damaging the original power (motor or turbine), this usually happens when the generator fails, and the machine operates as a synchronous motor connected to the power system. In this case, the generator becomes an active load, causing a rise in temperature and severe damage to the main turbine, and hence it becomes a need to quickly detect these conditions. This study proposes a new controller for Neuro-Fuzzy to prevent reverse power flow and to keep the quality and reliability of supply. Fuzzy system network has attracted various scientific and engineering researchers. The new feature of this work is to adjust the membership function as a reverse mechanism derived of the Fuzzy Logic Controller. The smart meter network is the basis of the smart grid. In this study, smart grid meters were implemented using ZigBee technology based on wireless sensor networks. The ZigBee network of wireless sensors due to its low battery, low power consumption, become more useful than other wireless communication systems to provide a high-performance measurement. This study shows the ZigBee network using the OPNET simulation. Depending on the performance, parameters were analysed to understand the operating characteristics of the star, tree, and mesh.

Keywords: smart grids, ZigBee IEEE 802.15.4, neuro-fuzzy network, directional relay

1. Summary

Literature reviews play a vital role in improving renewable energy because science is still a cumulative effort in the first place. As with any discipline, the synthesis of rigorous knowledge becomes indispensable to keeping up with the is growing searched pace of smart grid domain, which is developing exponentially by academics and engineers and scientific searched in the content of many papers, evaluated and synthesized [1–3]. The proposed study could provide a theoretical basis for confirming the need for investigative questions, proving that research methods have increased accumulated knowledge. Besides, high-quality reviews have made researchers look for a lot of literature when conducting empirical research.

In addition, high-quality reviews have made researchers look for a lot of literature when conducting empirical research. To conclude, our main objective in this chapter is to develop solutions to improve the spread of distributed energy and with high-speed synchronisation communication, that is central to the continuous development of the smart grid field. We hope that ours. This chapter will serve as a valuable source for those conducting, evaluating or engineers in this important and growing domain. The future distribution network may include a large-scale distributed power generation penetration into the smart grid. This scenario is aimed at the transition from a passive distribution network to an active distribution. The integration of DG units has a significant impact on the operation of power flow, voltage distribution, and protection systems in the distribution network [1]. (1) Explore issues that drive the demands of future rapid Intelligent protection systems, (2) design and develop a protection strategy that can be applied to any grid equipped DG. The concept of innovative protection must ensure the selectivity of protection in case of failure, (3) apply the new concept of intelligent protection algorithm.

As a result, the performance of the existing distribution network's traditional inverse-time protection system was evaluated. In this way, we have identified the problems faced by the current applicable protection strategies; the results of the simulation prove that the traditional protection system is insufficient to provide a satisfactory level of protection selectivity. This chapter introduces the transformation of the traditional protection strategy to the future intelligent distribution network protection system. This shows how unprecedented advances in sensor technology and the emergence of new communication protocols have stimulated innovation in protection systems. The latest technological advances have enabled existing protection systems based on local information to be transferred to innovative security systems, In Addition; the details of the new communication mechanism for the application of high-speed protection systems were discussed. Clever's protection strategies are fast, flexible and offer a high level of selectivity protection.

This chapter designs and develops a new concept of intelligent protection strategy. This approach applies to any network administered by DG. The proposed intelligent protection system aims to reduce the time to eliminate failures, to ensure the selectivity of protection and to enhance the availability of the units of the DG throughout faults. The new scheme of realising a protection scheme using advanced sensor, neural fuzzy scheme and ZigBee network is expounded. The intelligent algorithm ensures the selectivity of the protection by minimising the time of failure and eliminating the problem of the large time disconnection in the system [1].

2. Introduction

The smart grid (SG) is the next generation of power grids. Its purpose is to overcome the problems that exist in the conventional power grid. Smart grid technology has been used, such as sensors and communication networks, and advanced software and sensors to provide control and enhance the protection and optimisation of all network components, including production, transmission, and distribution.

Although neural networks implement to solve tuning problems, the fuzzy logic controller is intended to use structured knowledge in the form of rules [4, 5].

The combination of Fuzzy Logic and neural networks provides the ability to solve optimisation problems. This new method consolidates the established advantages of both approaches and avoids the limitations of both approaches. Control algorithms are used to prevent unexpected fluctuations in voltage and frequency. Smart grids use energy storage systems and communication networks to ensure total coordination between power generation and energy use. Reduce the energy loss of the network to minimise demand and energy costs [6–8]. A reliable, real-time information flow among parts of the network is critical to the success of the smart grid self-regulation process. There are many wireless standards for technical applications [9]. One of the most popular technologies is ZigBee wireless sensor network (WSN), which is distributed on the smart grid structure, which has a lot of equipment to communicate with each other through the wireless network. These electronic devices are called Sensors/Detectors/Transducers [9]. Sensors are devices that can recognise several of the physical units, such as current, voltage, impedance, etc. Also, the ZigBee system is featured by low energy consumption. It is also more economical than other communications because it provides flexibility and scalability [10–12].

The construction of this chapter is as follows: Section 2 confers a study of the SG communication system. The proposed protection system and results are analysed in Section 3; Section 4 the wireless sensor network using OPNET and simulation results; In Section 5, the conclusions are discussed.

3. The smart grid communication network architecture

The old communication system is characterised by limited efficiency and limited information exchange; the intelligent metering network is the backbone of a smart distribution network. It is essential to select the appropriate communications to facilitate the real-time flow of bi-directional information. It is mainly used by the main transmission point and a limited number of sensors on the transmission line for control and fault detection. Compared with the traditional network, an intelligent system contains a much larger number of sensors. Sensors are used to exchange information between terminals devices and data centres to handle such a large data stream, the SG must-have reliable communications and security infrastructure. The communication infrastructure must be self-managed and configured to change automatically [13].

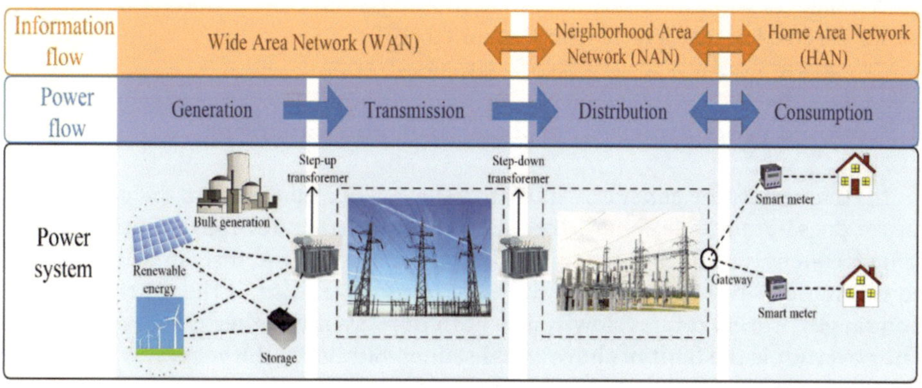

Figure 1.
Communication network architecture [14].

Figure 1 shows the smart grid communication architecture, including the neighbourhood network (NANs), home LAN (HANs), wide area network (WANs), substation and data centre. These networks as follows will be displayed briefly [14–16].

3.1 Wireless area network (WAN)

It serves as backbones that help the power grid that provides communication between utility systems and substation systems. The can help prevent power outages by providing real-time information from the electricity grid. It supports real-time control and protection. This system is useful when dealing with unforeseen contingencies, and it is essential to avoid interruptions and failures [17–20]. This application helps in performing a generator process and provides support for large power systems. The main disadvantage of this kind of WLAN is the possibility of devices interfering at the same frequency is high. The network operates between 2.4 and 3.5 GHz. Among its advantages are low-cost equipment, the use of which has spread across a wide range of applications.

3.2 Network home area (HAN)

Some technologies introduced in the Home Network System are ZigBee, WLAN with PLC. The construction of a Building Area Network (BAN) is considered to be more complicated than the Home Area Networks (HAN). The HAN can be classified as a part of the customer network structure; HAN is often used by consumers in the housing and business sectors, using power tools to communicate [21]. It is a combination of connected devices, management software and dedicated LAN. HAN supports communication between smart meters and appliances used in homes, industries or buildings. It supports several other services, including Demand Response, pre-payment, real-time pricing and load control. The essential of the HAN communication system is to include low-cost, low power consumption with secure communication [6, 10, 22].

3.3 Neighbourhood area network (NAN)

NAN is best described as a bridge between WAN and HAN and used in a NAN to collect data of points adjacent with the help of intelligent electronic devices (IEDs), which are widely deployed in the whole area. It is a two-way communication technology developed that give information about the control system for smart grids. Compared to WAN, the data rate is not high, and the transmission power is low for short-range transmission. WLAN, PLC and ZigBee are some of the techniques on which the NAN network can be implemented [23].

4. Proposed protection system simulation and modelling

In some cases, the generator starts to behave like a motor when the prime mover does not provide enough torque to keep the generator rotor rotating at the same frequency as the line of the parallel power source, and instead of giving power; it draws power from the parallel power source. Also, if the synchronisation ranges process rotates slowly, also both the loss of the alternator excitation. The governor is the fault of the original sender. Similarly, the generator will also extract the current from the source line [24]. When the rotating part of the generator fails, the generator stops generating electricity and starts drawing electricity from the parallel power source [25]. This situation may damage the drive unit and is not desirable. It should be detected as soon as a possible problem, and quickly disconnect the equipment from the parallel power supply, thereby protecting the generator from damage. In exceptional turbine cases, the power supply direction is changed from line to generator. It usually uses a directional protection relay to monitor the current flow and take appropriate action to prevent all outage case. The directional protection relay is working when the reverse power exceeds a certain percentage of the rated power output; it will trip the circuit breaker of the generator, disconnect the generator from the line under not normal circumstances, the relay setting is about 5% of the power generator [26]. The directional protection relay is located on the generator latch cabinet and is an integral part of the circuit breaker. The structure of the relay is designed to limit the reverse current flow depending on the amount of current and voltage between the two phase angles. If the line power is inverted, the current through the relay current coil will be inverted concerning the polarisation voltage and provide directional torque [27]. This technique compares the relative phase angle between the (current and voltage), as shown in **Figure 2**.

Typically, the phase angle is used to define the fault compares to the reference value. The voltage is usually applied as a reference amount. By comparing the operating voltage and current phase angle, can be inferred the fault occurs. Therefore, the fault current can be described with the phase relationship with the voltage line −90 for the forward fault, 90 for reverse fault. The relay wills response to the phase angle difference between the two quantities to come out trip signal [28]. In cases where optimal protection is required, Rogowski coil current sensors are used as CT and PT to avoid faults in conventional AC Transformers and must set a certain amount of delay during operation, to prevent power fluctuations, the transient effect during synchronisation. If the angle between the current vector and the voltage is Δ, the power flow is $-900 < \Delta < 900$ [29]. Under normal conditions, the voltage overlaps with the current range is more significant than their non-overlapping interval. However, in the case of reversing energy flow, this overlap is reduced to a lower level. **Figure 3** shows this assembly and implementation. The low signal of the current and voltage of RC sensors changes to form a square wave having a value of "±1" and then multiplying these level signals to produce a positive number in the overlap interval, in the negative numbers are generated in the non-overlap interval [30].

The integration limit of the scheme is set to zero, therefore the integration of the load is perpetually <0 under normal conditions. However, in the opposite trend, the production condition system as a whole tends to decline until the threshold constant

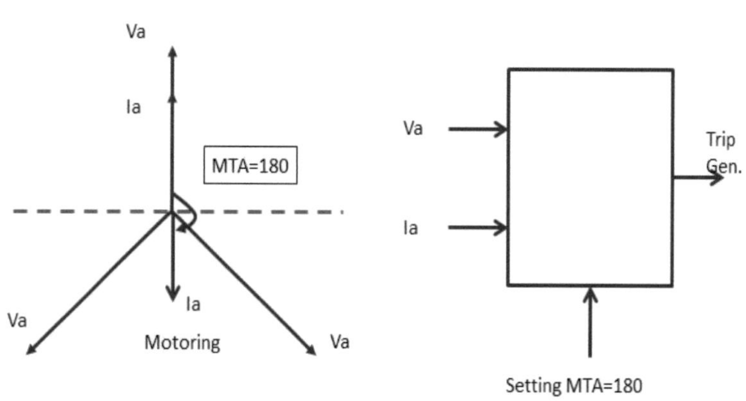

Figure 2.
According to the reverse power relay current with a voltage vector diagram.

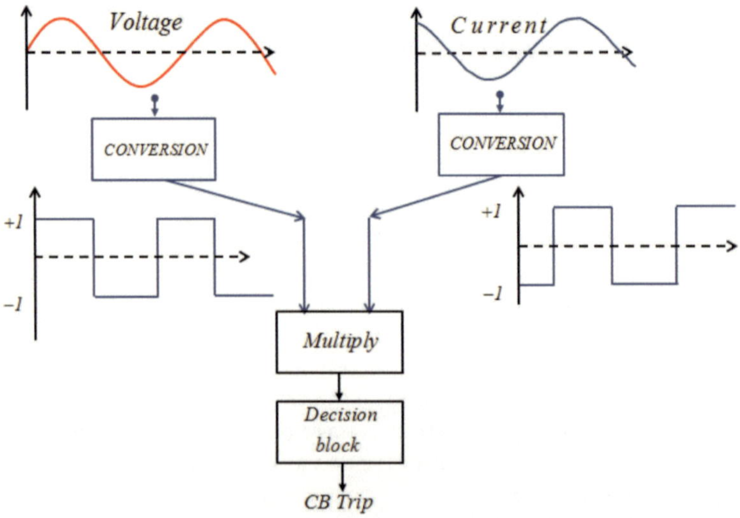

Figure 3.
Diagram of implementing a directional component.

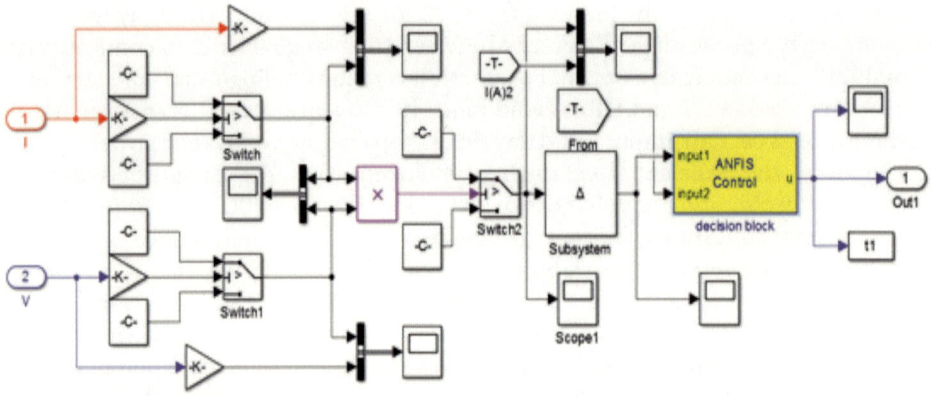

Figure 4.
Modelling of a directional relay component.

is reached. In this situation, the constant is set to 0.01 and the select value based on the amount of reverse power [23]. The output of the reverse power relay (RPR) is transferred to a decision where the production is one for normal operation, zero for abnormal conditions, as displayed in **Figure 4**.

Figure 5(a) presents the 3ø current directions, cos ø, and power factor, and **Figure 5(b)** the same ideas of the P and Q expansion.

4.1 The adaptive neuro-fuzzy approach

The selection of the membership function dramatically affects the quality of the fuzzy controller. Therefore, the method requires a more fuzzy logic controller. In this paper, a new method of neural networks is used to solve the adjustment problem of a fuzzy logic controller. We consider a dynamic system of multiple entrances, a single exit. The system is exported to the desired state of the control action can be described by the concept of the well-known “if-then” rule, where the input variables are first converted to their respective linguistic

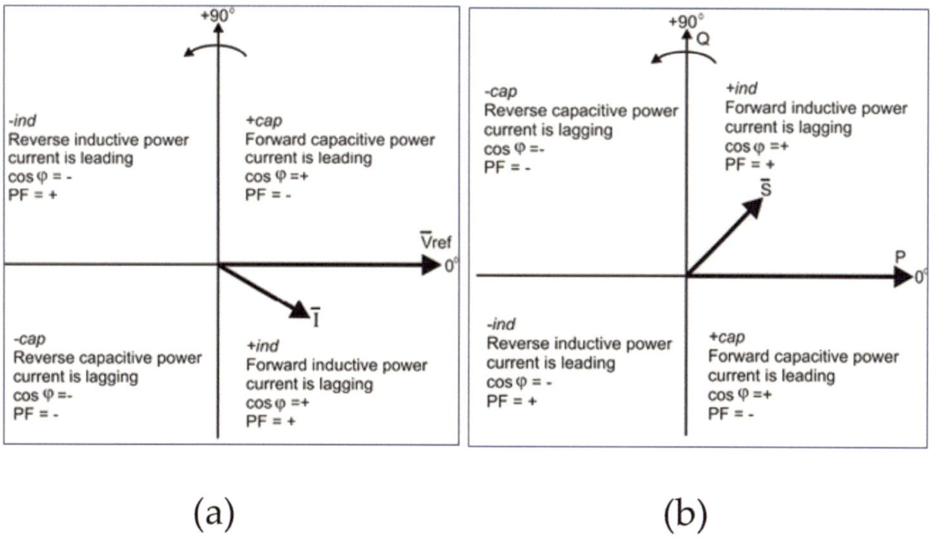

Figure 5.
(a) Quadrants of current/voltage. (b) Quadrants of a power.

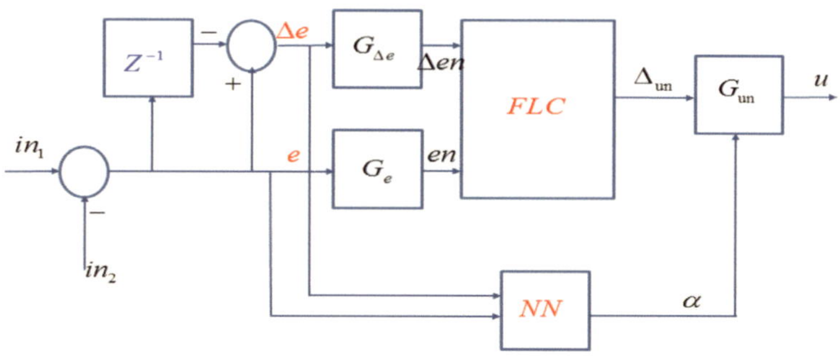

Figure 6.
Control method using adaptive neuro-fuzzy.

variables, also known as fuzzification. The value of these rules certainly output. Then use defuzzification to convert the output to a precise value. For simplicity, we used a modified centre of area method, and the Triangle fuzzy set will be used for input and output.

The linguistic form of the control rules is the basis of the designed fuzzy unit. It depends on the accuracy of the choice of parameters, which is the translation of the linguistic rules of the fuzzy set theory. The neural network (NN) is used to improve the selection of these parameters. In this scenario, the neural network is combined with the fuzzy logic unit. As shown in **Figure 6**, it uses the first fuzzy logic rule and then uses the neural network to generate the automatic adjustment output. References input [in (1)] related to the existing input [in (2)], product e(t), and incremental changes Δe(t) [31, 32]:

$$\Delta e(t) = e(t) - e(t-1) \qquad (1)$$

The proposed unit has two input factor gain measures of control, Ge and GΔe, and one scaling gain GΔu. The output-input scale factors are expressed as follows:

$$\Delta_{eN}(t) = \Delta e(t).G_{\Delta e}, \tag{2}$$

$$e_N(\mathrm{t}) = e(t).G_e \tag{3}$$

In the same eN and ΔeN scaling factor system to identify the fuzzy logic controller input signal product [33, 34]. Fuzzy logic controller output signal is ΔuN, it is the scale factor input. The neural network has two inputs, e(t) and Δe(t), and the neural network signal output α, which is used to fine-tune the product control of the operator. The output signal of the scale factor can be expressed by the formula:

$$\Delta u(t) = \Delta u_N(t)\alpha.G_{\Delta u} \tag{4}$$

The output signal can be written as follows:

$$u(t) = \Delta u(t) + u(t-1) \tag{5}$$

The results are displayed in **Figure 7**, which illustrates the specified fuzzy rules. We have selected fuzzy set and membership functions, **Table 1** summarizes the development of the rules used in this study [1].

Forming a neural network composed of three layers (two input layers, three hidden layers and one output layer). Neural network input (NN) is including the same number of output of fuzzy logic. The activated function has a value from −1 to +1 for the output signal, as shown in **Figure 8** [35–37].

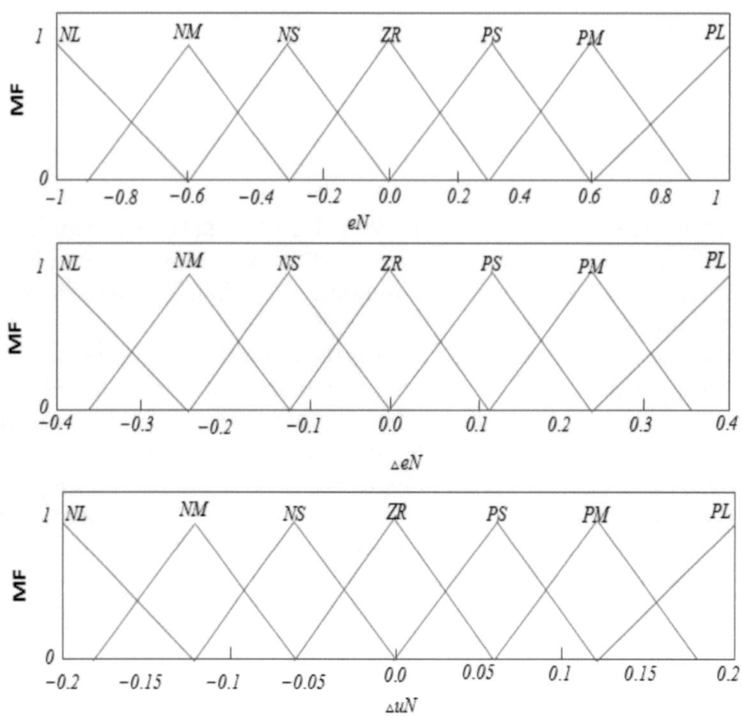

Figure 7.
Fuzzy logic membership functions.

Δe_N \ e_N	NL	NM	NS	ZR	PS	PM	PL
NL	PL	PL	PM	PM	PS	PS	ZR
NM	PL	PM	PM	PS	PS	ZR	NS
NS	PM	PM	PS	PS	ZR	NS	NS
ZR	PM	PS	PS	ZR	NS	NS	NM
PS	PS	PS	ZR	NS	NS	NM	NM
PM	PS	ZR	NS	NS	NM	NM	NL
PL	ZR	NS	NS	NM	NM	NL	NL

Table 1.
Rules of FL.

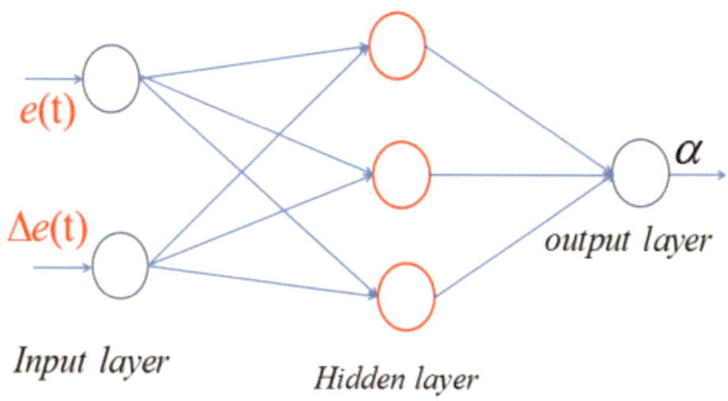

Figure 8.
The description of the neural network compositions.

$$f(x) = \frac{1 - e^{-x}}{1 + e^{-x}} \tag{6}$$

The activation function neuron in the output layer is:

$$h(x) = \frac{1}{1 + e^{-x}} \tag{7}$$

Control Unit based on the measured output signal $u(\text{t})$ from neuro-fuzzy circuit to adjust the trip circuit:

$$\theta_{new} = \theta_{initial} - k.u \tag{8}$$

where the θ initial is the initial switching output and k is a constant.

4.2 Simulation results

As shown in **Figure 9**, the simulation design uses 200 MVA /11 kV, with a synchronous generator connected to a transmission line 25 kV through an 11/25 transformer, 60 Hz, Load10 MW, and 3 Mvar. Relays are tested in a variety of situations. The conditions and results of the discussion are as follows.

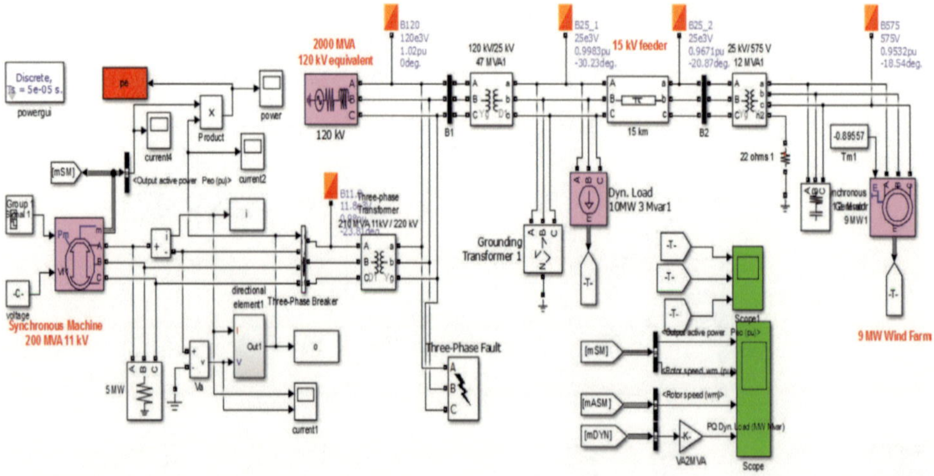

Figure 9.
Model of a reverse power relay in an electrical power system.

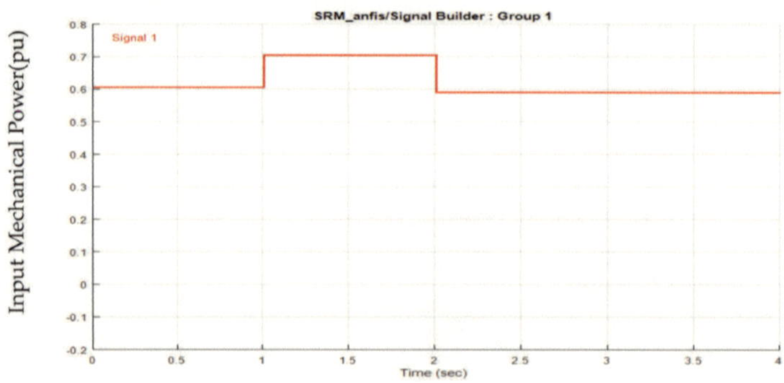

(a)

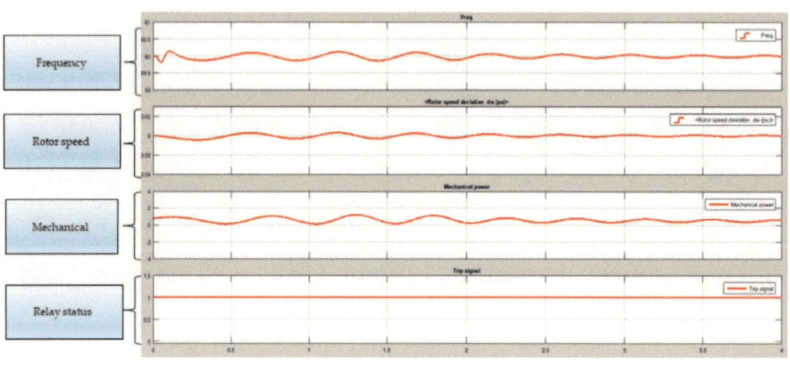

(b)

Figure 10.
(a) Performance of input-output power; and (b) relay status.

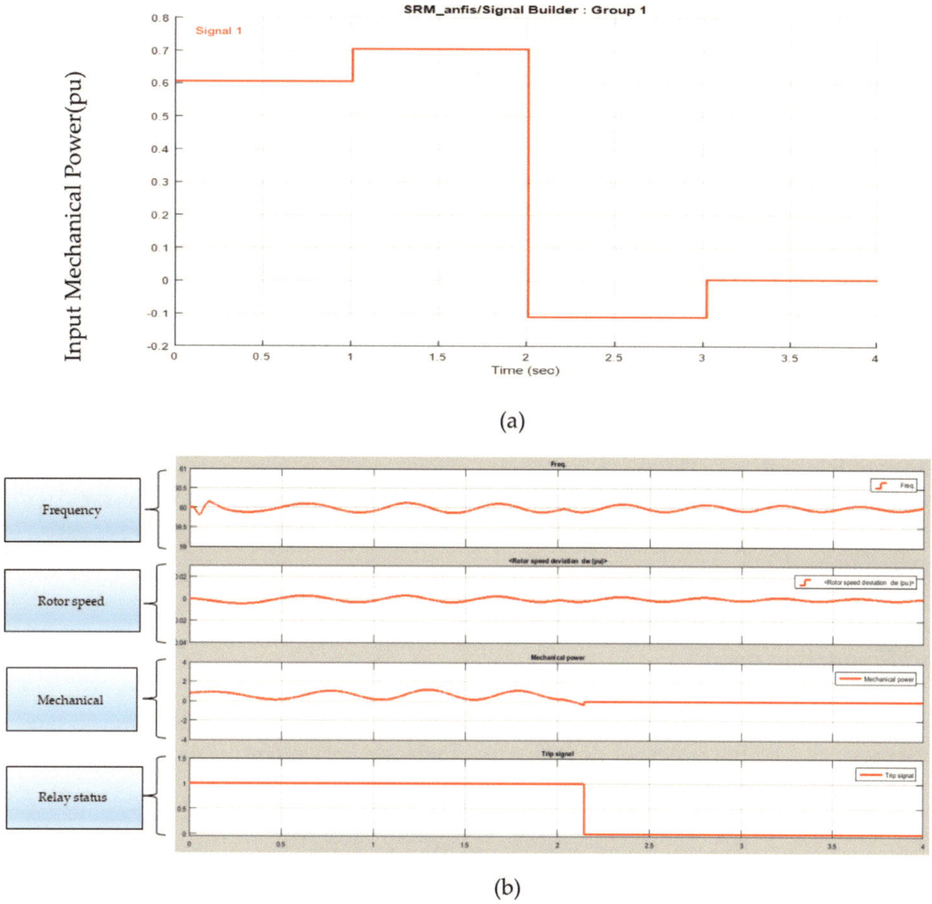

Figure 11.
(a) Performance of the input-output power; and (b) relay status.

The structure of propose protection is shown in **Figure 4**. The nodes of in_1 and in_2 represent the input variables and pass their values to the blocks that contain the respective membership functions in Neuro-Fuzzy controller. The relays are tested under a variety of conditions. We have provided the details of the system in **Table A1** [1].

4.2.1 Simulation results under the normal condition

In this case, the mechanical power input of the generator within 1–2 seconds differs from 0.6 to 0.7 pu, at the under normal circumstances the observed state is shown in **Figure 10**, and the relay does not trip.

4.2.2 Simulation results under the faulty condition

In this case, the mechanical power input in 2–3 seconds from 0.7 to −0.1 pu. Relay responds to this change after 0.15 second for safe, and the relay is triggered, where the fault occurred at 2 seconds as shown in **Figure 11**. Input Mechanical Power (pu).

The reverse current adjustment knob and the delay time are shown in **Table A2** of the Appendix, and then the trip is confirmed with the minimum reverse current in the range of 2–20%. The trip time delay setting range is 0–20 seconds.

5. Wireless sensor network using OPNET simulator

5.1 Zigbee network method

The OPNET modeller is one of the most important simulation tools for communication network inspection. ZigBee networks are known for their low power consumption, low cost, low data rate, and high battery life.

The current work as shown in **Figure 12**, consists of the workstation featuring a coordinated connection to six routers, with eight nodes installed at a range (200 meters) from each other. To participate in the calculation of system variables, the OPNET collected a large number of variables. The indicators relate to two types of statistical data for the agreement: local and global statistics. However, in terms of network performance, this study is more occupied in collecting quantitative information for the system. As a result, current research is based on data obtained from global statistics [1].

The values of the design parameters are shown in **Figure 13**, and the values of the parameters on the router are shown in **Figure 14**. Transmission power is estimated to be 0.1 w.

The ZigBee coordinator parameter is illustrated in **Figure 15**.

5.2 Simulation results

The simulation results are simulated under different topologies of the wireless sensor network, and the effects of different topologies on network efficiency are discussed.

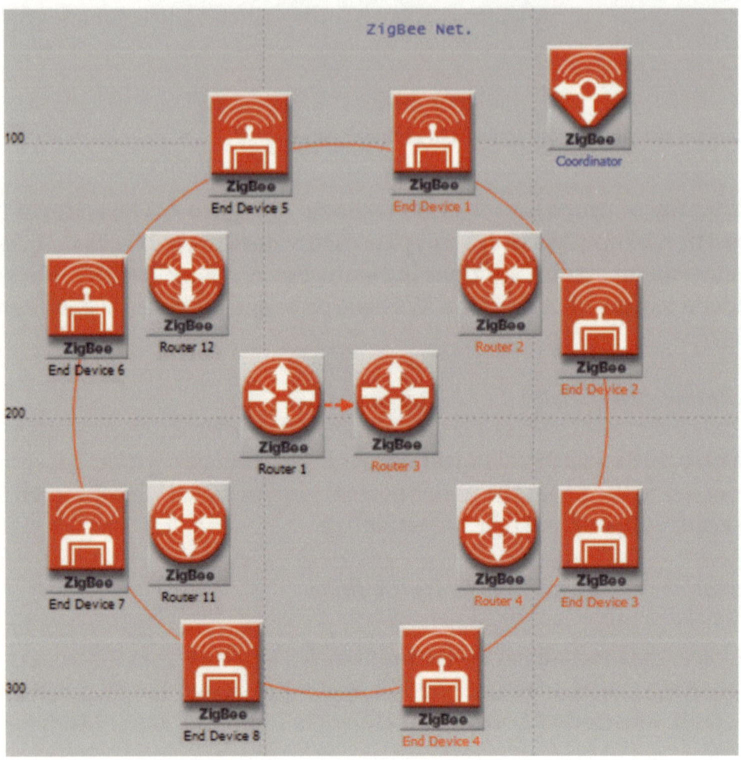

Figure 12.
Basic scenario consisting of one coordinator, (6) routers and (8) end devices.

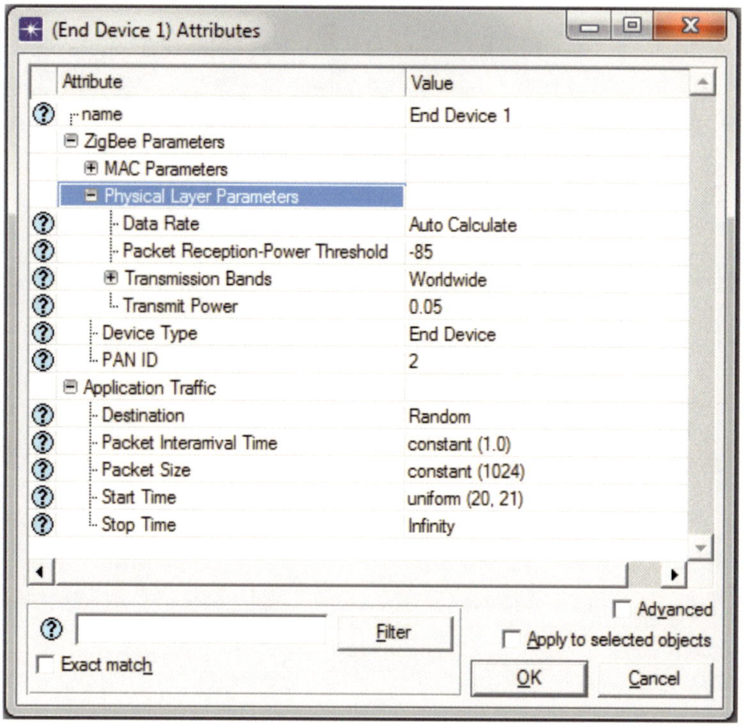

Figure 13.
End-device parameters.

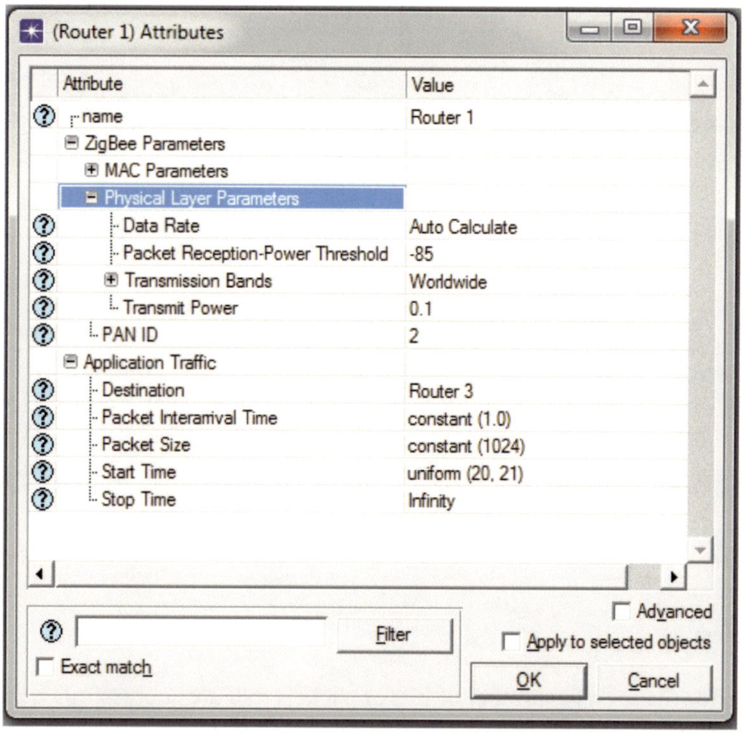

Figure 14.
Router parameters.

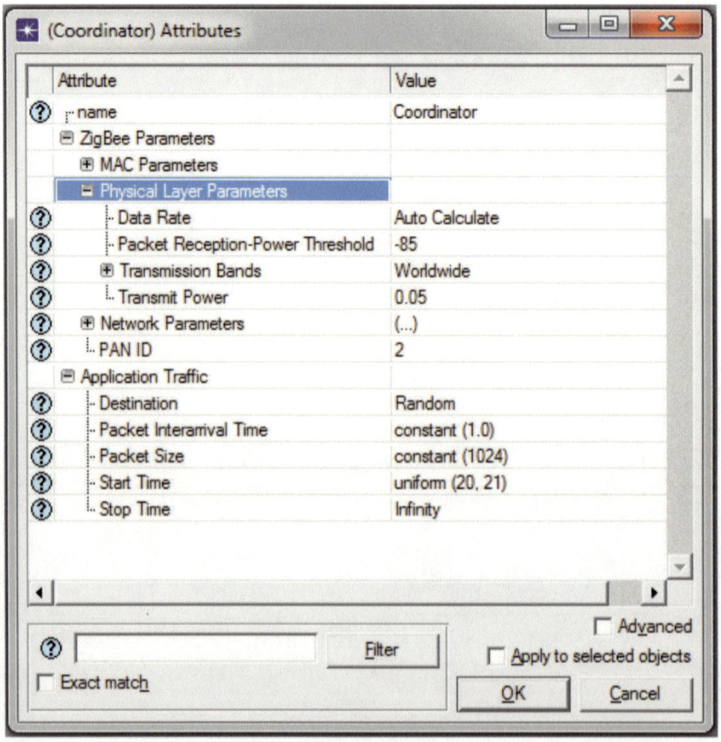

Figure 15.
ZigBee coordinator parameter.

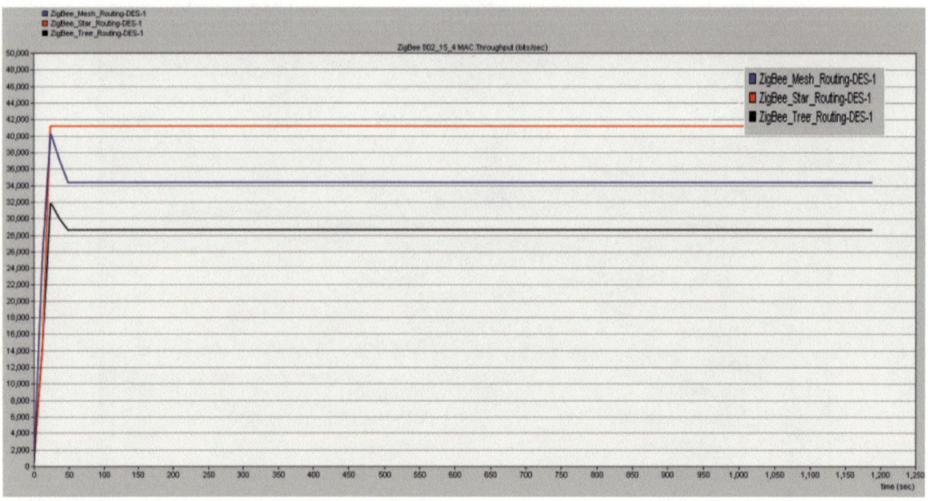

Figure 16.
Throughput case.

5.2.1 Throughput

Defined as the average number of bits or packets that are successfully transferred from source to destination. The steady-state results for the star, mesh and tree topology are 0.041, 0.034, and 0.028 Mbit/s, In the star topology, can achieve maximum throughput, this finding is that the star topology interacts with the personal area network (PAN) coordinator, (**Figure 16**).

5.2.2 Data traffic sent

As shown in **Figure 17**, and finding indicates that the maximum data traffic is in a star topology because this topology type allows communication with the coordinator. The data traffic sent was 0.1465, 0.0385, and 0.0325 Mbit/s, for a star, mesh and tree topology [1].

5.2.3 Data traffic received

Data traffic is defined as the number of data bits received per unit of time. **Figure 18** shows that the received data traffic for (star, mesh, and tree) topology is 0.650, 0.650, and 0.3805 Mbit/s.

This discovery means that the traffic received in the star topology is the largest because all devices communicate via the PAN coordinator and are responsible for generating traffic and routing [1].

Figure 17.
The total number of data bits transmitted.

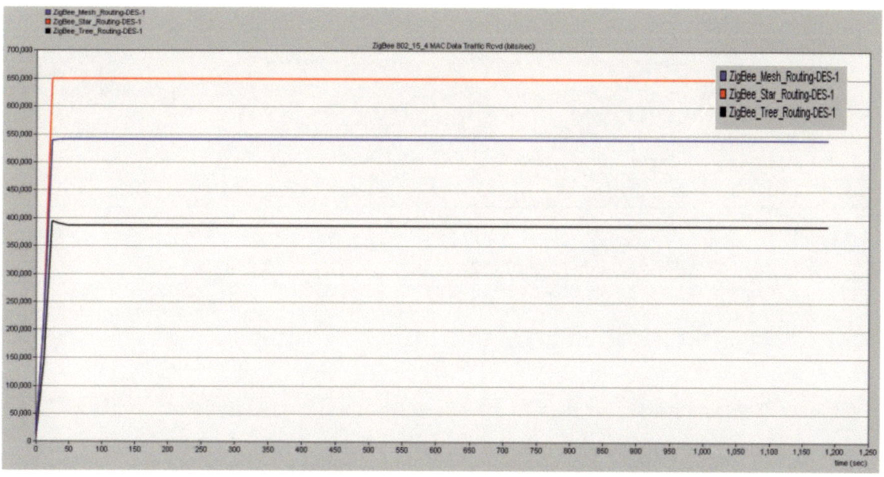

Figure 18.
Data traffic received.

5.2.4 End-to-end delay

It is the time it takes for a home target application to get the package generated by the source application. The results show the mesh/tree, and star topology delays are 9.6 and 7.9 ms. And the delay time of the mesh/tree topology is longer than that of the star topology, as shown in **Figure 19**. In a star topology, only one parent object is represented by a ZigBee coordinator. Therefore, the final mobility of the device may cause some delay.

5.2.5 Medium access control (MAC) load

As shown in **Figure 20**, MACload is used for forwarding the load for each PAN in the transmission of packets in the IEEE 802.15.4 MAC, that is, the physical layer, in the upper layers. The performance of the MACload presents similar results to the throughput performance. In other words, this result confirms the conclusion that

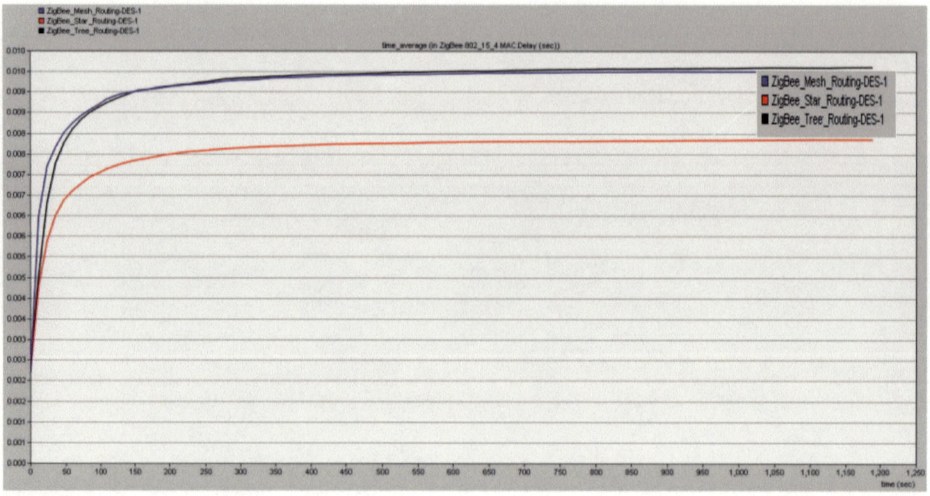

Figure 19.
Data arrival rate against delay.

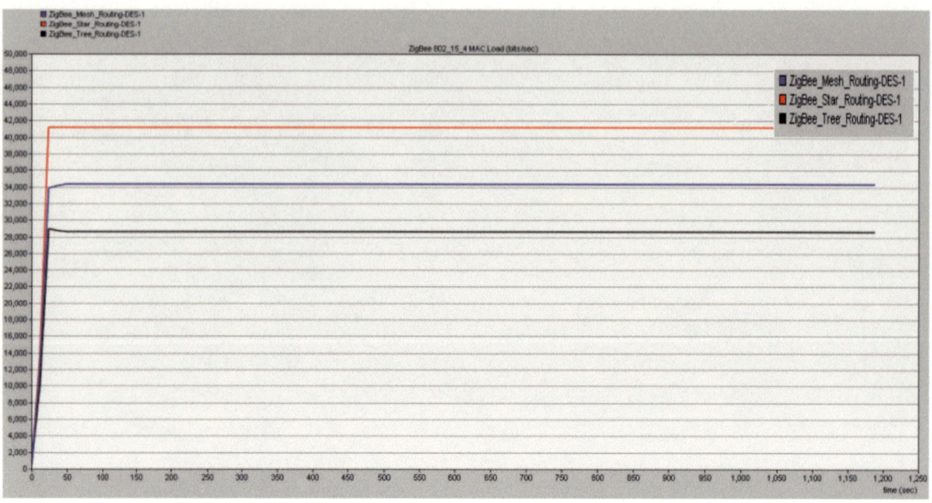

Figure 20.
Simulation scenario against a MAC load.

Definition	Value
Test zone (radius)	~100 meters
Number of end devices	8
Number of routers	6
Number of coordinators	1
Mobility model	Random
Simulation duration	1200 s

Table 2.
Summary of the simulation parameters.

the faster the load transfer to the upper layers from the physical level is, the more efficient the network. As shown in **Table 2** the local routing information covers only a small area (the diameter of the test distance is about 250 meters) [1].

6. Conclusions

The difference between the ZigBee and the WiMAX mobile networks is the distinction in their technology standard. The WiMAX mobile networks used in the simulation employ the IEEE 802.16 standard technology, whereas ZigBee follows the 802.15.4 standard. Mobile WiMAX seems to have better functionality than ZigBee, but taking into account the scalability of the latter, the former can install additional ZigBee devices because of its low-cost features and the possibility of reducing the battery size and operation hours. However, ZigBee may be more effective in certain areas because if its low energy consumption rate. The advantages of proposed protection are as follows:

(1) To prevent the flow in the opposite direction, and damage to the generator or the main engine. (2) To avoid the occurrence of explosion or fire, this is mostly caused by unburned fuel in the generator.

The existing power system is undergoing significant changes. Smart grid technology is the method used in the future power system framework, the integration of energy and communications infrastructure is inevitable. Intelligent network technology is characterised by the realisation of a complete dual communications infrastructure, automatic measurement, renewable energy integration, distribution automation and network monitoring. Wireless network to achieve the collection and transmission of real-time data. With flexibility in a wireless sensor network, high detection accuracy, low cost and excellent performance. Therefore, it can be used to develop interesting remote sensing applications. Implementation of sensor networks must meet the flexibility, scalability, cost, equipment, changes in the topology of the environment and energy consumption and other factors and limitations. Wireless sensor network has the flexibility, with high precision sensing, low cost and other excellent characteristics. Therefore, the sensor network must meet the flexibility, scalability, cost, environmental topology changes and energy consumption and other factors. The performance analysis of the topology of the ZigBee wireless network was carried out by using OPNET 14.5 simulators. The network topology of the star, tree and mesh is compared according to the end-to-end delay, throughput, Mac traffic load, and the four parameters of the transmit and receive traffic parameters. In terms of star topology throughput, the MacLoad is higher than the resulting value of the mesh topology, so the use of star topology is considered to be very important. The network types of the star, tree, and grid are compared according to the end-to-end delay, throughput, MAC traffic load, and the four parameters of the transmit and receive traffic parameters. The star topology is the best in terms of performance and has a MAC load that is similar to the mesh topology. Since the ZigBee network has a large number of nodes, so the use of star topology is considered to be very important.

Acknowledgements

The author would wish to thank the editors and critics for their constructive comments and suggestions. The project was supported by the BETC/Southern Technical University.

Author contributions

I would like to thank Ms. Janan Abd-Ali and Dr. Shaorong Wang for overseeing my studies. Thank you very much, Dear Ms. Sandra Maljavac, and all the staff who have cooperated with you, and I wish you all the best.

A. Appendix

Set point	Value
Range	~2–20% reverse current
Time delay	Adjustable 0–20 seconds

Table A1.
Technical data for reverse power monitoring.

Microgrid	Parameters	Value
Generator	Voltage	11 k V L-L, S = 200 MVA
Transformer	Voltage	V_P/V_S (L-L) = 11 kV/220 kV
	Frequency	60 Hz
Feeders	Line impedance	R = 0.02 X, L = 0.64 mH
	15 km feeder	R/km = 0.4 Ω, X/km = 0.3 Ω
Load	Dyn load	10 MW, 3 Mvar

Table A2.
Microgrid simulator parameters [1].

Author details

Ali Hadi Abdulwahid[1,2]

1 Engineering Technical College, Southern Technical University, Basra, Iraq

2 School of Electrical and Electronic Engineering, Huazhong University of Science and Technology, Wuhan, Hubei Province, China

*Address all correspondence to: dr.alhajji_ali@yahoo.com

References

[1] Abdulwahid AH, Wang S. A novel method of protection to prevent reverse power flow base on neuro-fuzzy network for smart grid. MDPI, Sustainability. 2018;**10**:1059. DOI: 10.3390/su10041059

[2] Abdulwahid A, Wang S. Application of differential protection technique of domestic solar photovoltaic based microgrid. International Journal of Control and Automation. 2016;**9**:371-386

[3] Khan MS, Woo M, et al. Smart city and smart tourism: A case of Dubai. Sustainability. 2017;**9**(12):2279. DOI: 10.3390/su9122279

[4] Baronti P, Pillai P, Chook VSC, Gotta SA, Hu YF. Wireless sensor networks: A survey on the state of the art and the 802.15.4 and ZigBee standards. Computer Communications. 2007;**30**:1655-1695

[5] Akram U, Khalid M, Shafiq S. Optimal sizing of a wind/solar/battery hybrid grid-connected microgrid system. IET Renewable Power Generation. 2018;**12**(1):72-80

[6] Leccese F. Remote-control system of high efficiency and intelligent street lighting using a ZigBee network of devices and sensors. IEEE Transactions on Power Delivery. 2013;**28**(1):21-28. DOI: 10.1109/TPWRD.2012.2212215. ISSN: 0885-8977

[7] Cagnetti M, Leccese F, Proietti A. Energy saving project for heating system with ZigBee wireless control network. In: 11th International Conference on Environment and Electrical Engineering; 2012. pp. 580-585. DOI: 10.1109/EEEIC.2012.6221443

[8] Mishra BC, Panda AS, Rout NK. A novel efficient design of intelligent street lighting monitoring system using ZigBee network of devices and sensors on embedded internet technology. In: 14th International Conference on Information Technology; 2015. pp. 200-205. DOI: 10.1109/ICIT.2015.37

[9] Kaleem Z, Ahmad I, Lee C. Smart and energy efficient led street light control system using ZigBee network. In: 12th International Conference on Frontiers of Information Technology; 2014. pp. 361-365. DOI: 10.1109/FIT.2014.74

[10] Qin H, Zhang W. ZigBee-assisted power saving for more efficient and sustainable ad hoc networks. IEEE Transactions on Wireless Communications. 2013;**12**(12):6180-6193. DOI: 10.1109/TW.2013.110813.130035

[11] Leccese F, Leonowicz Z. Intelligent wireless street lighting system. In: 11th International Conference on Environment and Electrical Engineering; 2012. pp. 958-961. DOI: 10.1109/EEEIC.2012.6221515

[12] Chen K-L, Chen Y-R, Tsai Y-P, Chen N. A novel wireless multifunctional electronic current transformer based on ZigBee-based communication. IEEE Transactions on Smart Grid. 2017;**8**(4):1888-1897

[13] Berger LT, Schwager A, Escudero-Garzás JJ. Power line communications for smart grid applications. Journal of Electrical and Computer Engineering. 2013;**2013**:1-16

[14] Kumar S, Bhattacharyya B, Gupta VK. Present and future energy scenario in India. Journal of The Institution of Engineers (India). 2014;**95**(3):247-254

[15] Maw HA, Xiao H, Christianson B. A survey of access control models in wireless sensor networks. Journal of Sensor and Actuator Networks.

2014;**3**(2):150-180. DOI: 10.3390/jsan3020150

[16] Callaway E, Gorday P, Hester L. Home networking with IEEE 802.15.4: A developing standard for low-rate wireless personal area networks. IEEE Communications Magazine. 2002;**40**:70-77

[17] Youn M, Lee J. Topology control algorithm considering antenna radiation pattern in three-dimensional wireless sensor networks. International Journal of Distributed Sensor Networks. 2014;**11**:1-11

[18] Leccese F, Cagnetti M, Trinca D. A Smart City application: A fully controlled street lighting isle based on raspberry-pi card, a ZigBee sensor network and WiMAX. Sensors. 2014;**14**(12):24408-24424. DOI: 10.3390/s141224408

[19] Pešović U, Mohorko J. Hidden node avoidance mechanism for IEEE 802.15 4 wireless sensor networks. Electronic Components and Materials. 2013;**43**:14-21

[20] Ajay ML. Fundamentals of Cellular Network Planning and Optimization. Canada: Willey-Interacience Publication; 2004

[21] Leccese F, Cagnetti M, Calogero A, Trinca D, di Pasquale S, Giarnetti S, et al. A new acquisition and imaging system for environmental measurements: An experience on the Italian cultural heritage. Sensors. 2014;**14**(5):9290-9312. DOI: 10.3390/s140509290

[22] Shi JF, Chen M, Yang Z. Power control and performance analysis for full-duplex relay-assisted D2D communication underlaying fifth-generation cellular networks. IET Communications. 2017;**11**(18):2729-2734

[23] Sharaf H, Zeineldin H, El-Saadany E. Protection coordination for microgrids with grid-connected and islanded capabilities using communication assisted dual setting directional overcurrent relays. IEEE Transactions on Smart Grid. 2018;**9**(1):143-151

[24] Ibrahim ME, Abd-Elhady AM, Power frequency AC. Voltage measurement based on double wound Rogowski coil. IET-High Voltage. 2017;**2**(2):129-135. DOI: 10.1049/hve.2016.0091

[25] Farjah E, Givi H, Ghanbari T. Application of an efficient Rogowski coil sensor for switch fault diagnosis and capacitor ESR monitoring in nonisolated single-switch DC-DC converters. IEEE Transactions on Power Electronics. 2017;**32**(2):1442-1456. DOI: 10.1109/TPEL. 2016.2552039

[26] Leccese F, Cagnetti M, Di Pasquale S, Giarnetti S, C aciotta M. A new power quality instrument based on raspberry-pi. Electronics. 2016;**5**:64. DOI: 10.3390/electronics5040064

[27] Abdulwahid A, Wang S. A Busbar differential protection based on fuzzy reasoning system and Rogowski-coil current sensor for microgrid. In: IEEE PES Asia-Pacific Power and Energy Engineering Conference. 2016. pp. 194-199. DOI: 10.1109/APPEEC.2016.7779496

[28] Ibrahim M, Abd-Elhady AM. Differential reconstruction method for Power frequency AC current measurement using Rogowski coil. IEEE Sensors. 2016;**16**(23):8420-8425. DOI: 10.1109/JSEN.2016.2614352

[29] Aman MM, Jasmon GB, and Khan QA, et al. Modeling and simulation of reverse power relay for generator protection. In: Power Engineering and Optimization Conference (PEDCO) Melaka, Malaysia, IEEE International. 2012. pp. 317-322

[30] Ehrenberger J, Svec J. Directional overcurrent relays coordination

problems in distributed generation systems. Energies. 2017;**10**(10):1452. DOI: 10.3390/en10101452

[31] Cervantes J, Salazar S, Chairez I. Takagi-Sugeno dynamic Neuro-fuzzy controller of uncertain nonlinear systems. IEEE Transactions on Fuzzy Systems. 2017;**25**(6):1601-1615

[32] Kermany SD, Joorabian M, Deilami S, Mohammad ASM. Hybrid islanding detection in microgrid with multiple connection points to smart grids using fuzzy-neural network. IEEE Transactions on Power Systems. 2017;**32**(4):2640-2651

[33] Abdulwahid A, Wang S. A novel approach for microgrid protection based upon combined ANFIS and Hilbert space-based Power setting. Energies. 2016;**9**(12):1042. DOI: 10.3390/en9121042

[34] Shahgoshtasbi D, Jamshidi MM. A new intelligent Neuro-fuzzy paradigm for energy-efficient homes. IEEE Systems Journal. 2014;**8**(2):664-673. DOI: 10.1109/JSYST.2013.2291943

[35] Lin WM, Hong CM, Cheng FS. Fuzzy neural network output maximization control for sensorless wind energy conversion system. Energy. 2015;**35**:592-601

[36] Garcia P, Garcia CA, Fernandez LM, Llorens F, Jurado F. ANFIS-based control of a grid-connected hybrid system integrating renewable energies, hydrogen and batteries. IEEE Transactions on Industrial Informatics. 2014;**10**:1107-1117

[37] Koraz Y, Gabbar HA. Fault diagnosis of micro energy grids using Bayesian belief network and adaptive neuro-fuzzy interference system. In: IEEE International Conference on Smart Energy Grid Engineering (SEGE); 2017. pp. 14-17. DOI: 10.1109/SEGE.2017.8052790